THE SWISS
TOUCH IN LANDSCAPE
ARCHITECTURE

景观设计中的瑞士印迹

[瑞士] 迈克尔·雅各布（Michael Jakob） 主编　翟俊 译

江苏凤凰科学技术出版社

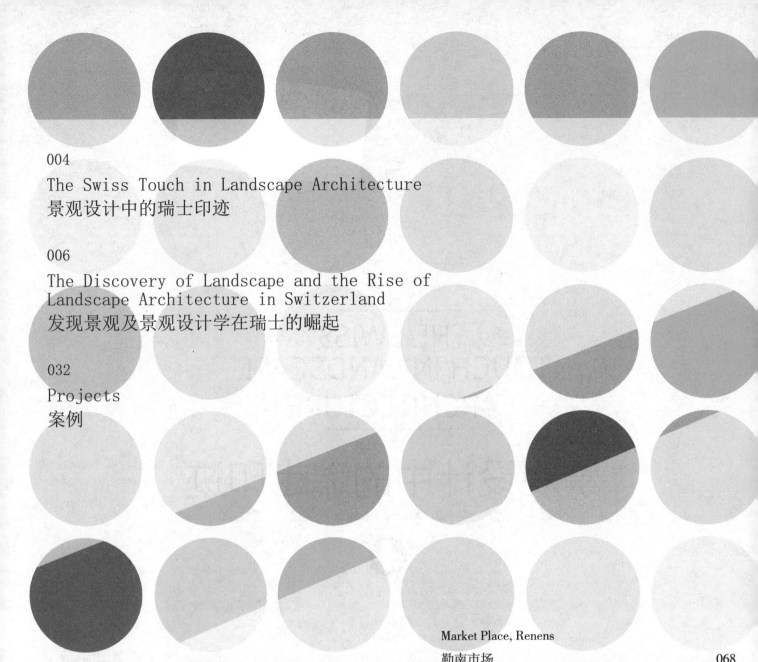

004
The Swiss Touch in Landscape Architecture
景观设计中的瑞士印迹

006
The Discovery of Landscape and the Rise of Landscape Architecture in Switzerland
发现景观及景观设计学在瑞士的崛起

032
Projects
案例

La Valsainte Charterhouse, Cerniat
La Valsainte 卡尔特修道院　　　034

Rudolf–Bednar Park, Vienna
维也纳鲁道夫 – 贝德纳公园　　　040

Eulach Park, Winterthur
温特图尔 Eulach 公园　　　044

The Tree Museum, Rapperswil–Jona
拉珀斯维尔 – 乔纳树博物馆　　　054

Prime Tower, Zurich
苏黎世泛利大厦　　　062

Market Place, Renens
勒南市场　　　068

Bear Park, Bern
伯尔尼熊公园　　　076

Letten Swimming Area, Zurich
苏黎世 Letten 游泳区　　　084

Ruggächern Residential Complex, Zurich
Ruggächern 商住综合楼　　　088

Spreebogen Park, Berlin
柏林施普雷河湾公园　　　098

Riedpark, Zug
楚格州里德公园　　　104

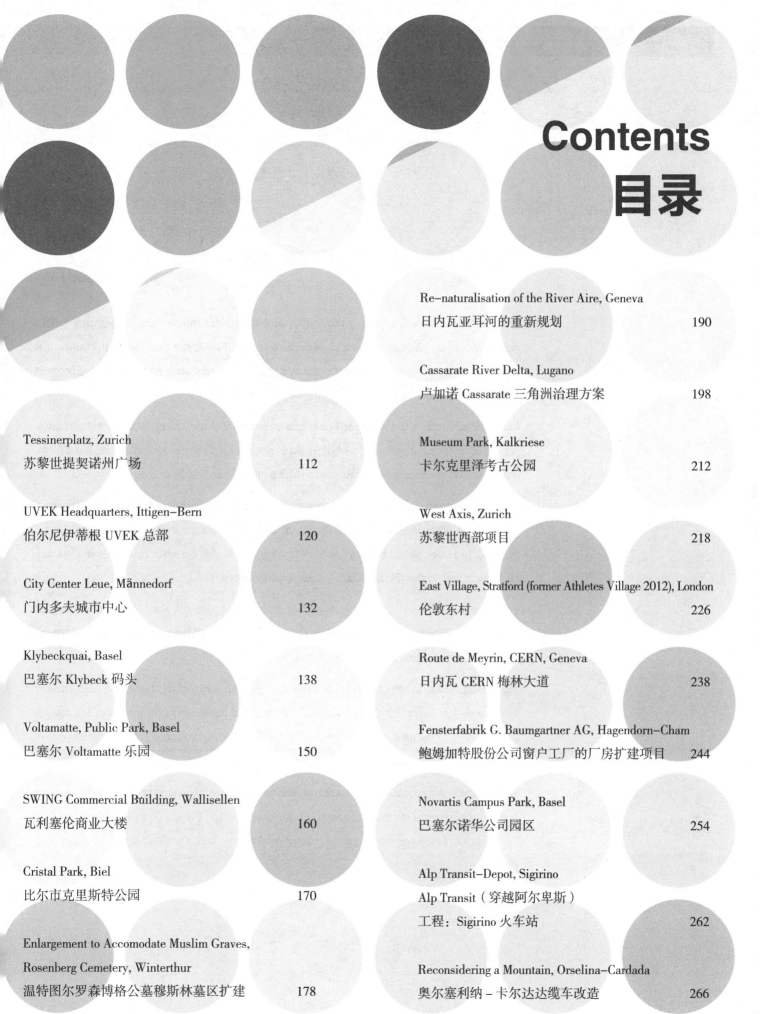

Contents 目录

Tessinerplatz, Zurich 苏黎世提契诺州广场	112
UVEK Headquarters, Ittigen–Bern 伯尔尼伊蒂根 UVEK 总部	120
City Center Leue, Männedorf 门内多夫城市中心	132
Klybeckquai, Basel 巴塞尔 Klybeck 码头	138
Voltamatte, Public Park, Basel 巴塞尔 Voltamatte 乐园	150
SWING Commercial Building, Wallisellen 瓦利塞伦商业大楼	160
Cristal Park, Biel 比尔市克里斯特公园	170
Enlargement to Accomodate Muslim Graves, Rosenberg Cemetery, Winterthur 温特图尔罗森博格公墓穆斯林墓区扩建	178
Re-naturalisation of the River Aire, Geneva 日内瓦亚耳河的重新规划	190
Cassarate River Delta, Lugano 卢加诺 Cassarate 三角洲治理方案	198
Museum Park, Kalkriese 卡尔克里泽考古公园	212
West Axis, Zurich 苏黎世西部项目	218
East Village, Stratford (former Athletes Village 2012), London 伦敦东村	226
Route de Meyrin, CERN, Geneva 日内瓦 CERN 梅林大道	238
Fensterfabrik G. Baumgartner AG, Hagendorn–Cham 鲍姆加特股份公司窗户工厂的厂房扩建项目	244
Novartis Campus Park, Basel 巴塞尔诺华公司园区	254
Alp Transit–Depot, Sigirino Alp Transit（穿越阿尔卑斯） 工程：Sigirino 火车站	262
Reconsidering a Mountain, Orselina–Cardada 奥尔塞利纳 – 卡尔达达缆车改造	266

The Swiss Touch in Landscape Architecture

Michael Jakob

Landscape architecture continually transforms the environment and the way we approach it. This modern discipline, which began around 1800, quickly evolved throughout the twentieth-century. It is now undergoing an even more spectacular rise as it takes a prominent place within society and becomes established as an important concern.

Landscape architecture is now focused on the development of public spaces, in urban and peri-urban projects. It concerns the management of green spaces within the city and the creation of parks and gardens. It also deals with the integration of waste land and agricultural terrain into the landscape.

During the twentieth century and beyond, Switzerland played an essential role in the evolution of landscape architecture. Leading projects by Swiss landscape architects can be found both in Switzerland and the rest of the world. Landscape architects play a central role among those professions which create present and future living spaces, and yet that role is still largely unknown to the general public.

The projects of Swiss landscape architects privilege formal beauty, but also place emphasis on ecological context. The need to respect both the spirit of a place and its history yields designs which combine, in equal measure, an extreme sense of rigour with elegance.

The present exhibition hints at the richness, diversity and complexity of this fascinating field. The first section deals with theory and history and provides a conceptual framework for the exhibition. The visitor is invited to discover the influence of the history of gardens and the important role played by the pioneers of Swiss landscape architecture.

The visitor will also see the crucial function of national exhibitions, including the more recent Lausanne Jardins – the successful international festival of urban garden design. The second section – the core of the Swiss Touch in Landscape Architecture – provides works of the most significant landscape architecture studios.

景观设计中的瑞士印迹

迈克尔·雅各布

景观设计学正在不断地改变环境以及我们对待环境的方式。这个现代学科大约起始于1800年，并在整个20世纪迅速发展。现如今景观设计更是经历着突飞猛进的发展，随着其社会地位的不断提升，已成为一项重要的课题。

景观设计当前注重市区及城市周边公共空间的开发，关注城市内绿色空间的管理及公园和花园的创造，同时还把荒地和农业用地与景观相融合。

20世纪以来，瑞士在景观设计的发展中起到了至关重要的作用。在瑞士以及世界的其他地方，瑞士景观设计师的杰作随处可见。景观设计师在为现在和未来创造生活空间的职业中起着核心作用，但是，这一重要性仍未被大众所广泛认知。

瑞士景观设计师的创作追求形式之美，但是同样重视和强调生态意识。因为尊重地方精神和当地历史，这使得他们的设计作品兼具严谨与优雅的极致感。

本书所展示的案例表现了这个迷人领域的丰富性、多样性和复杂性。第一部分为相关理论和历史，为本书提供了概念框架，使读者了解景观设计的影响以及瑞士景观设计学先驱所起到的重要作用。

读者还将看到国家展览的重要作用，包括最近的洛桑园艺展——有关城市园林设计的国际盛事。第二部分是本书的核心内容——景观设计中的瑞士印迹，介绍了最重要的景观设计工作室的作品。

The Discovery of Landscape and the Rise of Landscape Architecture in Switzerland

发现景观及景观设计学在瑞士的崛起

Michael Jakob（迈克尔·雅各布）

图1

图2

In Europe, landscape is a relatively recent invention. Originally, from the 15th to the 18th century, landscape meant exclusively landscape painting. European landscape begins somewhere around 1350, with Ambrogio Lorenzetti's significant *Good Government fresco* (ill. 1), in Siena, Italy. This complex painting already shows that landscape – that which exists outside the city walls, extra muros – is an invention of the city, of urban intelligence, and that it is mastered by it (landscape, here the "contado", is under the control of and belongs to the city and to its techniques of representation). Technically speaking, the Siena wall painting is not exactly a landscape, but still the sum of topographical elements. Only from the 15th century on can we speak of landscape in the full sense: a piece of land represented from a precise point of view. Whether this happened first in Italy or in the Netherlands is still the object of controversy. More important to our subject is the fact that quite early in landscape painting history, mountains became a part of the new genre. We find them, in fact, in Leonardo da Vinci (his earliest work, a drawing from 1473 (ill. 2), shows a hilly region, while later drawings represent the Alps or fantastic landscapes) and in Joachim Patinir's paintings (ill. 3). We approach Swiss territory, that is, the contemporary territory of Switzerland, with the first recognizable landscape painting: Konrad Witz's the miraculous drought of the fishes (ill. 4). It shows a precise landscape in the western part of Switzerland: the city of Geneva, Lake Geneva, and two mountainous forms: the Salève and the Môle.

在欧洲，景观的发展历史相对较短。就起源而论，从15世纪至18世纪，景观专指景观绘画。欧洲景观始于约1350年的一幅非常重要的壁画，即意大利锡耶纳安布罗乔·洛伦泽蒂（Ambrogio Lorenzetti）的《好政府的寓言》（图1）。这幅复杂的绘画作品已经表明景观虽然存在于城市围墙之外，却是城市和城市智慧的产物（景观既是城市和技术的产物，亦受两者的限制）。从学术角度来说，锡耶纳的壁画不能算真正意义上的景观，它只是地形元素的综合。在进入15世纪以后，我们才迎来真正意义上的景观：一块土地某一视点的表现图。而其发源地是在意大利还是荷兰仍是争论的话题。但是，对我们讨论的主题更为重要的是在景观绘画历史早期，山脉成了新题材的一部分。事实上，在达芬奇的绘画（他最早期的作品，1473年的一幅绘画（图2），对山区进行了描绘，后来的画作描绘了阿尔卑斯山或者幻想景观）和约阿希姆·帕提尼尔（Joachim Patinir）的绘画（图3）中我们可以发现对山脉的描绘。 康拉德·维茨（Konrad Witz）的《基督履海》又名《捕鱼的奇迹》（图4），是第一幅能够清晰地辨认瑞士的景观绘画。这幅画描绘了瑞士西部的景观：日内瓦城、日内瓦湖以及萨雷布山和Môle山。

图3

What we recognize in Witz's work is a painted 'Swiss' landscape, and it would take a long time before the real experience of landscape, the possibility to enjoy, on the spot, a piece of nature became possible. In Europe, it took centuries for people to learn how to interpret and gain access to Nature, and only once they acquired the mental categories enabling them to frame Nature did they actually create landscapes. The process took such a long time for a fundamental reason: in order to stir fear no longer and be enjoyed as an aesthetic phenomenon, Nature had to be interpreted in a positive way. But such an attitude was impossible as long as Nature was seen as something globally negative. It was considered negative because, for the most influential Christian ideologies, Nature in its totality was deeply marked by human sin.

图4

从维茨的作品中，我们认识了画出来的瑞士景观，但是真正体验景观，并把自然作为景观来欣赏，还是经历了很长时间的等待。在欧洲，人们花费了几个世纪的时间来学习如何解读和走进自然，只有他们学会如何在头脑中对自然进行分类之后，才能真正地创造景观。这个过程十分漫长，其根本原因在于，为了不再畏惧自然，并把它当成一种美学现象来欣赏，人们首先应该以积极的方式解读自然。 但是，在当时普遍把自然看作消极事物的观念下，以积极的态度解读自然是根本不可能的。之所以认为自然是消极的，是由当时盛行的基督教意识形态决定的，按照这种意识形态，自然是人类罪恶的载体。

First of all, Adam and Eve had, once they ate from the "Tree of Knowledge", to leave paradise and live in an imperfect, cruel and sometimes gruesome Nature. The second 'event', the Great Flood, was even more important, because well into the 17th century the official theology insisted that we actually live in a postdiluvian world. Before the Deluge (ill. 5), the Earth had "not a Wrinkle, Scar or Fracture in all its body, no Rocks nor Mountains, no hollow Caves, nor gaping Channels, but even and uniform all over", as Thomas Burnet stated in his famous *Sacred Theory of the Earth* (ill. 6). But afterwards, mankind had to face a hostile environment (the "natura lapsa", Nature marked by the Fall) and to resist Nature, as far as possible. Everything linked to the senses, especially to sexuality, was considered a permanent source of danger for more than a millennium, which clearly explains why the mere fact of observing Nature was interpreted as a capital sin. A worldview with such an imprint consequently considered that untamed and wild Nature was full of diabolic forces, like dragons, snakes, basilisks and other fantastic beings, all embodying the Devil.

图 5

首先，亚当和夏娃吃下智慧树的果实后，不得不离开伊甸园，生活在严酷、恶劣甚至阴森恐怖的自然中。第二个事件大洪水之说也起着更为重要的作用，因为进入 17 世纪后，官方神学坚持认为我们实际上生活在大洪水后的世界里。在大洪水（图 5）之前，地球上无任何褶皱、疤痕或破裂，也无任何岩石、山脉、洞穴或峡谷，如托马斯·伯内特（Thomas Burnet）在他著名的《神圣的地球理论》（图 6）一书中所描述的那样，它是平整且统一的。但是后来，人类不得不面对恶劣的环境（亚当和夏娃堕落以后），不得不竭尽所能地抵抗自然。在相当长的一段时期内任何与七情六欲有关的，尤其是性欲，都被认为是罪恶的永久来源，这也解释了人们为什么把自然解读为罪恶源这一事实。在这种世界观的烙印下，不能被驯服且野性十足的自然被认为充满着魔鬼般的力量，如龙、蛇、蜥蜴以及其他怪物，它们都是恶魔的化身。

图 6

As long as people trembled before Nature (which they hardly observed at all) and as long as they thought that it represented the empire of diabolic forces, access to it as something worthwhile looking at or beautiful was a matter of impossibility. The same Swiss Alps that almost everybody began to admire and wished to experience at first hand from the 1750s on, were regarded until the 18th century as the territory of Evil. Only very few exceptionally enlightened individuals already understood in the second half of the 16th century that the most mysterious territory of Switzerland, the Alps, was full of positive "wonders" of Nature, of useful plants and other resources – and that the fresh air you could breathe there was a source of good health. Conrad Gesner (1516 – 1565), a famous Zurich physician, scientist and humanist (ill. 7) was such an exception.

只要人类在自然面前感到战栗恐惧（人们几乎觉察不到），只要他们认为自然是魔鬼的化身，那么把自然作为值得观察或有美感的事物去接近几乎是不可能的。自18世纪中期起几乎人人称赞并渴望体验的瑞士阿尔卑斯山仍被认为是邪恶之地。在16世纪后期，只有极少数进步人士把充满神秘色彩的瑞士领土如阿尔卑斯山，看作是积极的自然"奇观"，认识到这里充满有价值的植物以及其他资源，有可以呼吸的、有益健康的新鲜空气。康拉德 • 格斯纳（Conrad Gesner）（1516—1565），著名的苏黎世医生、科学家和人类学者（图7），就是其中的代表之一。

图7

As early as 1541, Gesner, the "German Pliny", a pioneer in various fields, including zoology, botany and plant geography, had written to his friend Vogel (in his preface to Libellus de lacte et operibus lactariis) about the healthy effects of the mountains and, in 1555, he even climbed one of the peaks of the Pilatus. Another famous Swiss scholar, Johann Jakob Scheuchzer (1672 – 1733), shows how powerful the idea of the biblical Flood still was at the beginning of the 18th century. Scheuchzer (ill. 8), a physician, mathematician and prolific author, discussed in his *Itinera alpina tria* (London, 1708) the presence of dragons in the Lucerne canton, attested by men of "good faith". Although he doubted the value of such testimonies (ill. 9), he added a series of illustrations of the terrible dragons. Furthermore, he was convinced that the fossils, an important object for his studies (as for Gesner), were without any doubt remnants of the Great Flood.

图8

早在1541年，格斯纳作为多个学科领域的先驱，包括动物学、植物学和植物地理学，在给朋友沃格尔的书信中就提到了山脉的健康效用，1555年，他甚至登上了皮拉图斯山的一座顶峰。而另一位著名的瑞士学者约翰 • 雅各布 • 居泽尔（Johann Jakob Scheuchzer）（1672—1733），则表明进入18世纪时强大的圣经大洪水思想仍然存在着。居泽尔（图8）作为一名医师、数学家和多产作家，在所著的 *Itinera alpina tria*（伦敦，1708）一书中讨论了卢塞恩州龙的存在性，以及"虔诚"人士的证词。虽然对这些证词的价值存在怀疑，他还是在书中增加了一系列可怕的龙的插图。此外，他确信，他所研究的重要对象——化石，毫无疑问是大洪水的残迹。

The necessary and fundamental element in gaining access to nature – a process that at a certain moment gave rise to landscape, that is, the framing of certain places and the pleasure linked to it – is culture. Culture creates Nature. There is, in other words, nothing more artificial or manmade than Nature. The desire or love for Nature is therefore never natural or given: it is rather the result of complex and long cultural processes.

促使人们走进自然（走进自然这一过程在特定阶段导致了景观的产生，也就是说，只有进入自然的一些区域后，才引发人们对自然的兴趣）的必要和基本要素无疑是文化。 文化创造了自然。换言之，没有任何事物比自然更具人造性或人塑性。对自然的渴望或热爱从来都不是天生的或赋予的，而是复杂而漫长的文化作用的结果。

The fact that, globally speaking, 'Swiss Nature', and especially Alpine Nature, became the object of (European) desire during the 18th century, is the result of several historical factors. The first one was Physicotheology, a very influential philosophical and scientific theory between 1650 and 1730. In opposition to the traditional Decay-explanation, Physicotheology stated the existence of divine "Design". Marked both by Anglican Protestantism and by the New Sciences and their mechanistic worldview (the world as a perfect machine), it explained that Nature was not negative, adverse to man and ugly but, on the contrary, the mirror of divine order. The new religious idea of a principally positive Nature gave way to the encounter with her and actively invited everyone to study even its smallest details and mechanisms. Physicotheology acted in this way as a prime motor pushing people to "go out" into even the wildest Nature, exploring and representing it in both words and by images. Starting in Britain, the physicotheologic movement spread quickly on the continent where it contributed greatly to a new reading of Nature altogether. However, for a long time, to recognize and to study Nature didn't imply any aesthetic attitude. The positive quality of the natural phenomena was the result of their utility and not of their beauty.

图 9

广泛而言，"瑞士自然"，尤其是阿尔卑斯自然，在 18 世纪成为（欧洲）渴望之地是多种历史因素共同作用产生的结果。第一个是自然神论，它是 1650 至 1730 年之间非常有影响力的哲学和科学理论。与传统的解释相反，自然神论说明了神圣"设计"的存在。以英国新教徒、新科学以及他们的机械论世界观（把世界作为一个完美的机器）为代表，他们认为自然不是消极的，不是不利于人类和丑陋的，相反，它是神圣秩序的反映。积极自然新的、主要的宗教观念使其与自然的本质相契合，积极邀请人们研究其最微小的细节和机制。自然神学以类似于原动机的方式促使人们"走出去"，进入并探索最野性的自然，并用语言和图画的方式把它表现出来。起源于英国的自然神学运动很快就传播到整个欧洲大陆，这对重新解读自然做出了巨大贡献。但是，在很长时间以内，这种对自然的认识和研究并未涉及任何美学上的态度。对自然现象的积极解读是出于它的实用性，而与美丽或美学层面无关。

Both the new optimistic worldview and commercial interests drove more and more people into even the remotest regions. In order to prove divine providence and design, every single corner of Nature had to be studied carefully. The wild Nature of the Alps didn't yet please the eye, but it provided pleasure to the intellect. In this regard, the presence of artists in the Swiss Alps is very revealing. The first artist who studied and depicted the Swiss mountains, the Dutch painter and etcher Jan Hackaert (ill. 10), did so because commissioned by a wealthy lawyer from Amsterdam, Laurens Van der Hem.

新的乐观主义世界观和商业利益驱使着越来越多的人开始探索甚至是最遥远的地区。为了证明自然是神的安排和设计，它的每一个角落都被细致地研究。阿尔卑斯山的野性自然本性不仅悦目，而且还能带来心智上的愉悦感。关于这一点，瑞士阿尔卑斯山地区的艺术家对此进行了很好的揭示。首位研究并描绘瑞士山脉的是一名荷兰画家和蚀刻师 Jan Hackaert（图10），他这样做是因为受到了阿姆斯特丹一位富有律师 Laurens Van der Hem 的雇佣。

图 10

Hackaert worked between 1653 and 1665 in the eastern part of the Swiss Alps. Whether he was a "spy" and his admirable topographic drawings really served commercial purposes – to use the Via Mala as a particularly practical way of crossing the Alps – or whether he only contributed to the topographical Atlas of the powerful Van der Hem, the interest to study these remote places was indirect and mediated neither by scientific nor by artistic reason. One century later, at a time when the European Alpo- and Swissomania were already fully developed, the Swiss painter Caspar Wolf (ill. 11) discovered the mountains of his own country, motivated by external factors.

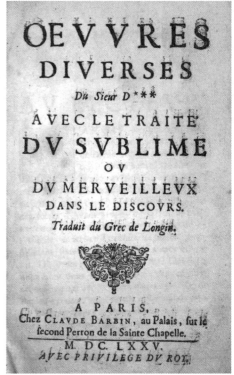

图 11

1653 至 1665 年，Hackaert 在瑞士阿尔卑斯山的东部开展工作。不管他是不是一个"间谍"，其令人钦佩的地形绘画能力的确为商业目的的活动提供了便利——把维亚玛拉峡谷作为跨越阿尔卑斯山的切实可行路线，或者不管他是否只为律师的地图集做出了贡献，可以证实的是研究这些偏远地区的兴趣并非只出于科学或艺术的目的。一个世纪以后，当欧洲 Alpo and Swissomania 已经被完全开发以后，受外界因素的激发，瑞士画家卡斯帕·沃尔夫（图 11）才发现了自己国家的山脉。

Only after his stay in Paris, and influenced by the great interest in Alpine landscapes in the French Salons of the 1770, did Wolf come back to Switzerland and start a journey across the Bernese Alps in particular. The result of his discoveries were commercialized by an art dealer in Bern, Abraham Wagner, who had a stock of 150 Wolf-landscapes; and disseminated in Europe, thanks to a famous series of ten etchings of the Lauterbrunnen valley. At the time when Wolf came on the landscape painting scene, a third factor had

图 12

already changed the representation of Nature, both artistic and mental. This factor consisted, once Nature was interpreted positively, in the aesthetics of the sublime. The new category of the sublime really transformed the way people experienced Nature because it permitted them, for the first time ever, to look at it in a different way. The sublime, originally a category of antique rhetoric and almost completely forgotten until the 17th century, was rediscovered by the French critic Boileau (ill. 12) and applied primarily to art. The fundamental step was the transposition of the religious and artistic sense of the sublime to Nature itself. It was accomplished by a rather obscure Englishman, John Dennis, a playwright and a critic who, while crossing the Alps during his Grand Tour of 1688, applied the sublime to the stunning scenery of impending rocks, precipices, torrents and peaks he encountered. Dennis had actually read both Burnet and Boileau, and the combination of these two sources with the scenery he discovered on site gave birth to an entirely new experience: "The sense of all this produced different motions in me, viz. a delightful Horror, a terrible Joy, and at the same time that I was infinitely pleased, I trembled."

1770 年，沃尔夫在巴黎居住，当时法国绘画沙龙对阿尔卑斯景观表现出极大的兴趣，受此影响他返回了瑞士，并开始了横跨伯尔尼阿尔卑斯山脉的旅行。他发现的成果被伯尔尼的一位美术经纪人变成了商品，这个人就是亚伯拉罕·瓦格纳（Abraham Wagner），他保存了 150 幅沃尔夫的景观画作，其中关于卢达本纳山谷的 10 幅蚀刻版画在欧洲广受传播。在沃尔夫回来开始景观绘画创作时，有一种因素已经改变了对自然的态度，无论是艺术上还是心智上，这种因素就是自然的崇高美学。这发生在对自然进行积极解读之后。这种新的分类的崇高之处在于其实实在在地改变了人们体验自然的方式，因为这使人们可以首次以不同的方式看待自然。崇高，这一起源于古代修辞学的词语，直到 17 世纪几乎被完全遗忘，却被法国评论家波瓦洛（Boileau）（图 12）重新发现，并首次应用于艺术中。最初只是把宗教和艺术上的崇高意义转移到自然上。而实现这一转移的是一个不太知名的英国剧作家和评论家约翰·丹尼斯（John Dennis），他在 1688 年的修学旅行中跨越了阿尔卑斯山，并开始用崇高一词形容他所遇到的岩石、悬崖、急流和山峰等令人震撼的风景。丹尼斯实际上读过伯内特和波瓦洛的著作，把这两人著作中的信息与他所发现的风景结合后，他产生了一种全新的体验："所有这些所带来的感官体验在我脑海中产生了不同的效应，有令人愉快的震惊和极度的喜悦感，同时我被无限的愉悦包围着，我颤抖了！"

Burnet and Dennis's travels were only the starting point of a series of Tours by intellectuals who experienced wild Nature and reacted to it in their writings. Joseph Addison – he did his Grand Tour in 1699 – is particularly important in this context. In his much read essay in the *Spectator* on the *Pleasures of Imagination* (1712), he was the first to identify natural grandeur (the sublime) as a legitimate source of a form of pleasure very different from the beautiful. In his reflections he even put forward a third intermediate term, newness, promised to an important future. The application of the sublime to the mountains helped to resolve a dilemma that persisted even after the rise of physicotheology. Although the idea of design functioned well for most of the forms of Nature, it didn't fit with the sense of terrible disorder experienced in the remote Alpine valleys. It was the sublime that permitted the formidable disorder of the Alps to be interpreted as something interesting and really 'natural' to look at. Once this idea became popular, pioneering travelers set out to tour the continent and the British Isles in search of these new kind of objects. Meanwhile, the explanations of the sublime became more and more sophisticated. Edmund Burke provided one of the first theories in his *A Philosophical Inquiry into the Origin of Our Ideas of the Sublime and Beautiful* (1757), where he stated: "Whatever is fitted in any sort to excite the ideas of pain and danger, that is to say, whatever is in any sort terrible, or is conversant about terrible objects, or operates in a manner analogous to terror, is a source of the sublime." While with Shaftesbury and Kant the sublime advanced to an even more central position in aesthetic theory, a growing number of people tried to be touched by these objects on the terrain or by reading contemporary novels full of sublime adventures.

伯内特和丹尼斯的旅行只是有识之士一系列旅行的出发点，他们开始体验野性的自然并在他们的作品中反映出来。关于这一点，约瑟夫·艾迪生——他于1699年开始修学旅行，表现得尤为突出。他在阅读量极大的《旁观者》上发表的小品文《想象的乐趣》（1712）最先把自然的壮丽（崇高）确定为乐趣的一种正当来源，而这种愉悦感与美感有着本质的不同。在他的想法中，他甚至提出了另一个中间术语——"新"，用于形容自然，这在未来对自然的解读起着很重要的作用。使用崇高形容山脉帮助解决了甚至是在自然神学兴起后仍存在的一种两难境地。虽然自然是一种神圣设计和秩序这种观念对于自然的大部分形态适用性很好，但是它与在偏远高山河谷所经历的、令人可怕的无序感并不匹配。 正是崇高一词的使用使人们可以把阿尔卑斯山这种可怕的无序解释为一种有趣的且看起来很真实的"自然"现象。一旦这种观念流行起来，先驱旅行者们便开始探索欧洲大陆和不列颠诸岛，寻求新的类似事物。与此同时，崇高这一解释变得越来越复杂。埃德蒙·伯克（Edmund Burke）在他《论崇高与美丽概念起源的哲学探究》一书中提出了这样一种理论："任何能够激起痛苦和危险的事物都是合适的，也就是说，无论任何可怕的、与可怕有关的事物或是以类似于恐怖的方式存在的事物都是崇高的起源。"而对于沙夫莎伯星（Shaftesbury）和康德（Kant）来说，崇高上升到了美学理论中更中心的位置，越来越多的人试着去实地或通过阅读充满崇高探险内容的现代小说来接触这些地貌特征。

图 13

At the beginning of the 18th century, the first guide to Switzerland (ill. 13) , Abraham Ruchat's *Les Délices de la Suisse* (Leiden, 1714), had to admit in his preface that speaking of "delights" in Switzerland sounded indeed quite strange: "Et depuis quand trouve-t-on des Délices en Suisse?" Especially for a foreigner, whom the anonymous author promises a "sincere and honest description, such a title must have been quite a surprise: " Les étrangers, qui ne connaissent notre pays que par les froides plaisanteries que l'on fait parmi eux, s'imaginent que c'est un pays de loup garoux, où l'on ne voit le soleil que par un trou; que ce ne sont que montagnes à perte de vue, que rochers stériles, que précipices affreux; que les habitants ne sont que des misérable vachers." By the end of the same century, and Johann Gottfried Ebels' Instructions pour un voyageur qui se propose de parcourir la Suisse (French edition, 1795; German first edition, 1787), everything had changed.

18 世纪初，作为首部瑞士旅行指南，Abraham Ruchat 在 *Les Délices de la Suisse* （Leiden，1714）一书的序言中不得不承认："在瑞士谈到"愉悦"，听起来确实很陌生，人们什么时候在瑞士发现了乐趣？"尤其是对于一名外国人来说，能够认识到瑞士所带来的愉悦感，这相当惊奇，"仅仅通过冷笑话了解我国的外国人以为瑞士是一个孤僻的国度，人们不了解外边的世界，并且认为在瑞士只有一望无际的山脉"（1795 年法国版，1787 年德国初版），其实一切早已经改变了 。

Ebel could now pretend "qu'aucune partie de notre Globe n'est aussi remarquable et aussi intéressante que la Suisse" and that both "l'homme et le philosophe" can make more observations and have purer pleasures in Switzerland than anywhere else in the world. Philosophers, scientists and those interested in politics could, in Switzerland, study the 'Great Book of Nature' (ill. 14) , the rules of Mankind (anthropology) and of Democracy, while enjoying the sublime spectacle of Nature. Switzerland, "the garden of Europe", had become the center of the world: "Whatever is grand, extraordinary, wonderful or sublime; all that can inspire us with fear and terror; all the bold, gloomy, and melancholy features, which Nature is pleased to exhibit in her compositions; whatever she displays in her immensity of romantic pleasing, peaceable, and pastoral scenes, seem to be collected in that country, to make it the garden of Europe … he who, in his rambles through Switzerland, has not been able to view Nature in the most propitious moments, cannot form a correct idea of her grandeur and enchanting sublimity. Inexhaustible in her forms, she exhibits everywhere new charms and new wonders; everywhere she shows herself to the eyes of the wondering observer under a new aspect. Amidst such a wild and sublime scenery, how many objects are there calculated to develop the resources of a poetical genius! How many interesting prospects invite the landscape-painter!" (The Traveler's Guide Through Switzerland)

埃贝尔（Ebel）现在可以声称"我们生活的地方任何部分都与瑞士有着联系"，景观和哲学家带来了更多的观察，世界上没有任何地方能够比瑞士更能给人带来愉悦感。在欣赏自然的崇高景色时，哲学家、科学家甚至是只对政治感兴趣的人都可以在瑞士研究自然这本大书和人类规则（人类学）以及民主主义。瑞士这一"欧洲花园"已经成为世界的中心："任何雄伟、非凡、惊人或崇高的事物，任何给我们带来恐惧和害怕的事物；自然所乐意展现的任何粗犷、黑暗、忧郁的特性，她给人带来的无限浪漫、舒适、平静和田园场景，所有这些似乎都在这个国家集中体现，这使得瑞士成为欧洲的花园　　在瑞士漫游时，既不能把自然视为吉祥之地，也不能对她的壮丽和使人入迷的崇高形成一种正确的观念。她变化无穷，到处弥漫着迷人的魅力和新的奇观；在充满惊奇的观察者眼中，她把自己各个新的方面展现出来。在如此野性、崇高的风景中，大部分是天才诗人灵感的来源，这些迷人的景色同样吸引着景观画家！"

图 14

This radical turn is due to many reasons and, generally speaking, to the accomplishments of travelers (ill. 15), poets, painters, scientists, philosophers, etc. However, two Swiss personalities had a major role in putting Switzerland at the center of European attention. The first one, Albrecht von Haller, was not only a celebrated scientist but, at least in his younger years, an important poet. After an excursion from Geneva to Zurich across the Alps in July and August 1728, Haller published a long descriptive poem, *Die Alpen*, in 1731, which rapidly became a European bestseller. Haller mixed physicotheological elements with the ideas of enlightenment, Protestantism and baroque visions of the Infinite. *Die Alpen* provided a reading of the alpine regions of his country as a sort of exotic paradise, protected from decadence:

Ye sons of Nature! still with you abide

Those goodly days; for 'mid your barren soil,

Estrang'd from tinsel vanity and pride,

Want is your happiness, your pleasure toil:

图 15

Such fair effects to man does virtue bring;

And though in frozen clouds your thirst you slake;

Though tedious winters nip the tardy spring,

And chilling snows your valley's ne'er forsake;

Yet does the savage clime your bliss increase,

While manners pure from guilt mark all your days with peace.

Then praise high Heav'n, that to your land denied

Riches, true source of ev'ry vice and ill;

While torrents wait on luxury and pride,

The heart of unaspiring want is still.

Wood were her temples, pulse her warrior's feast,

When Rome from ev'ry war triumphant came;

At length when wholesome moderation ceas'd,

Weak was her arm, her glory but a name:

While pure simplicity and temp'rance reign,

Oh bless your happy lot, nor pant for cursed gain.

(Edward Hamley, transl.)

这样一种根本性的转变是由多方面原因造成的，一般而言，是旅行者（图 15）、诗人、画家、科学家和哲学家等的成就造就了这一转变。但是，两名瑞士名人在把瑞士推向欧洲注意力的中心时起着重要作用。其中一个是阿尔布莱克·冯·哈勒，他不仅是一位著名的科学家，在他年轻的时候，还是一位重要的诗人。1728 年 7 月和 8 月，经历了从日内瓦到苏黎世，横跨阿尔卑斯山的远足后，哈勒出版了长篇描述性诗文《阿尔卑斯山》，1731 年，它迅速成为欧洲的畅销书。哈勒把自然神学元素与启蒙观念、新教教义和无限的巴洛克式想象力结合到一起。他所著的《阿尔卑斯山》一书中，他把阿尔卑斯山地区描绘成了一个免于衰落的异域乐园：

你的儿子自然，仍与你们同在

那些美好的日子，在你贫瘠的土地

从浮华、虚荣到骄傲

要的是你的幸福和快乐

美德带给人类的这种公平的影响

在寒冷的云中，你的欲望被平息

单调乏味的冬天之后是迟来的春天

也不放弃令人战栗的雪山峡谷

天气越恶劣，你就越狂喜

因内疚而变得有礼貌，记录着平静的日子

于是向上帝致敬，也向你否定的土地致敬

每个缺点和不幸的来源都是不一样的

洪流正高傲地等待

而那颗无抱负的心却心静如水

树木是她的神殿

当每次战争胜利的号角吹响时

勇士在此欢庆

当一切停止

她的手臂不再强壮，只留下一个名字

充满野性的自然

铭记与你在一起的平静日子

（爱德华·哈姆利　翻译）

Haller was the first major figure to popularize the beauty of the mountains, while advancing the idea of the honesty and proud sense of liberty of the local mountain dwellers. In his poem, he developed a new poetic language in order to name the variety of things seen in the Alps. Haller moved his readers to tears and was quickly translated into many languages.

哈勒是第一个在宣传山脉的美丽的同时还推崇当地山区居民的诚实和自由、淳朴的人。在他的诗文中，为了命名他在阿尔卑斯山脉所见的各种事物，他甚至开创了一种新型诗体语言。哈勒把他的读者感动到流泪，他的诗文也很快被翻译成多种语言。

Another book was still more important in putting Switzerland at the center of European attention: Jean-Jacques Rousseau's *La Nouvelle Héloïse*. This epistolary novel, published first with the subtitle "Letters from two lovers living in a small town at the foot of the Alps" (1761), was one of the most successful books of the 18th century. Rousseau (ill. 16, ill. 17) had a preference for places where he could calmly botanize and, thanks to his hero Saint-Preux, identified the region of the High Valais as an extraordinary place. The key ideas put forward in Saint-Preux' letters, such as the beauty of the secluded sites, the healthy impact of fresh air and the general grandeur of the Alpine scenery were at that time already well known topoi. What Jean-Jacques added to the enthusiastic description of this Swiss idyll was the subjective element. This unique Nature had, in other words, a direct influence on the feelings of Saint-Preux and, thanks to thousands of Rousseau-readers, on all those who tried to repeat their literary experiences by visiting the major sites of the Nouvelle Héloïse (ill. 18). From the 1760s on, Clarens, Vevey, Chillon (ill. 19), Meillerie and many other places around Lake Geneva were part of a literary circuit. In these years, Switzerland became a fashionable place (the laborious and virtuous Swiss mountain peasant replaced the passive shepherd of Sicily or Greece), especially in the intellectual Salons of the European capitals, and people flocked to admire – at first hand – the torrents, precipices, chalets, and views reminding them of the sentimental story of Saint-Preux and Julie.

图 16

图 17

在把瑞士推向欧洲注意力中心的过程中，另一本书的作用则更为重要。这就是让－雅克·卢梭的《新爱洛伊斯》。这是一部书信体小说，首次出版时的副标题为"居住在阿尔卑斯山麓的一个小城中的两个情人的书信"（1761年），它是18世纪最成功的著作之一。卢梭（图16，图17）特别喜欢那些能够让他安静地研究植物的地方，感谢小说的男主角圣·布莱尤思（Saint-Preux）把瓦莱州地区选作这一非凡之地。圣·布莱尤思的书信中所提出的主要观点，如幽静之地的美丽、新鲜空气的健康效果以及阿尔卑斯山风景的壮丽,在当时已经非常有名。在描述瑞士田园般景色的同时,卢梭还加入了主观元素。 换言之,这些独特的自然景观对圣 • 布莱尤思的感情产生了直接影响,成千上万的卢梭读者都想通过参观《新爱洛伊斯》中描述的地区(图18)再次体验在文学作品中的经历。自18世纪60年代以来,克莱伦斯(Clarens)、韦威(Vevey)、夏兰(Chillon)(图19)、梅耶里(Meillerie)以及日内瓦湖周围的许多其他地区成了文学范畴的一部分。在当时,瑞士成了最流行的地区(勤劳善良的瑞士山区农民替代了西西里岛或希腊的牧羊人),尤其是在欧洲首都知识分子沙龙中,人们蜂拥地去欣赏和赞美能够使他们想起圣 • 布莱尤思和朱莉之间伤感故事的湍流、悬崖、瑞士的农舍以及其他景观。

图18

Protestantism, physicotheology, pre-romantic sentimentalism, a certain disbelief in the values of enlightenment, the criticism of decadent civilization, a sense for disorder as a higher form of order, the interest in ruins and picturesque objects generally, the idea of freedom and democracy and, last but not least, the power of fashion based on well established topoi glorified Switzerland and set it up as the ideal place to study and to 'feel' Nature.

新教教义、自然神学、烂漫情怀、对启蒙价值观的某种不信任、对堕落文明之说的批判、把无序作为一种更高形式的有序、对废墟和如画般事物的兴趣、自由和民主主义思想,以及最后但同样重要的由赞美瑞士景观衍生出的对流行事物的推崇,所有这些为研究和"感受"自然奠定了思想基础。

However, around 1800, the situation was already quite complex. If, on the one hand, Swiss landscape occupied the highest place in the collective imagination of the time; on the other, several critical factors and voices had already begun to complicate its triumph. The hypertrophic circulation of guide books, travel literature, journals, etchings and other forms of representation made these places so incredibly popular that some, fatigued by too much Swiss imagery, decided to seek out less well known sites. The crisis of landscape painting and its tendency to distance itself from Nature was another major factor of disturbance. Rousseau's description of the Valais as "a real theater" proved quite exact at a moment when the first effects of mass tourism pushed too many people at the same time to the same places. During these years, when Switzerland topped the aesthetic and touristic hierarchy, there were already other 'Switzerlands' at hand: one new way to discover the country at the heart of Europe consisted simply in visiting the fashionable panoramas (after giving an overview of the cities, they quickly represented the crags of the Swiss Alps and other mountain scenes). Another way to enter the Alpine

图19

图 20

world was provided by panoramic wallpapers and scenic landscapes representing the Swiss Alps. One could thus 'travel' to Switzerland in one's own home, in a public building (the panorama) or in one's garden, where Swiss scenery was extremely popular too.

但是在 1800 年，形势已经变得相当复杂。一方面，瑞士景观在当时集体想象力方面占据最高地位；另一方面，个别批判性因素和声音已经开始把这种胜利复杂化。旅行指南、旅行文学、日志、蚀刻版画以及其他表现形式作品的过度发行使这些地方达到了空前流行的程度，以至于那些对瑞士形象感到疲惫的人决定到瑞士以外寻找那些不太著名的地方。景观绘画危机以及其远离自然这种趋势是导致形势变复杂的另外一个因素。卢梭把瓦莱州描述成"一个真正的大剧院"恰恰反映了当时太多人在同一时间趋之若鹜地前往同一地点所造成的结果。 在当时瑞士登上美学和旅游等级体系顶峰时，还有其他方式可以体验瑞士，一种新方法是通过简单参观流行的全景画（画中不仅有对城市的概述，还有对瑞士阿尔卑斯山以及其他山区景观的呈现）去发现这个位于欧洲中心的国家。进入阿尔卑斯世界的另外一种方式是通过全景壁画。这样，人们在自己家里、在公共建筑（全景图）或自己的花园里就可以"游遍"瑞士，这些地方的瑞士风景也流行到了极致。

The relation between Switzerland as an aesthetic construction and the so called English or picturesque garden of the 18th century is particularly interesting. Let us not forget that, in advocating a taste for the irregular and natural, Addison's *Spectator* essay on the *Pleasures of Imagination* discusses an aesthetic form applicable to both savage Nature and the new natural gardens. The same sentimental longing for Nature stands at the origin of the interest in the wild scenery of the Alps (ill. 20) and the natural scenery built in the landscape gardens (ill. 21) of the first half of the 18th century. When Addison argues that "in the wide Fields of Nature, the Sight wanders up and down without Confinement, and is fed with an infinite variety of Images, without any certain Stint or Number", we can easily apply this to both natural scenery and the picturesque gardens of the immediate future. But there is more in question here than a simple analogy between Alpine disorder and variety on the one hand, and the irregular forms created in the landscape gardens of Kent, Brown, Bridgeman and other designers on the other. Switzerland, "the garden of Europe", is not only present in these places through topography, but as an intertext. In presenting torrents and waterfalls, the landscape gardens and parks of the 18th century literally imitated the Alpine landscape. Another frequent element of the time were the extremely popular Swiss features, for instance the Swiss bridge or the Swiss chalet. By citing Switzerland, these far away places in Britain or on the continent became metonymically a part of the model they imitated.

瑞士（作为审美的构筑物）与所谓的 18 世纪英式或画境园林之间的关系非常有趣。我们不要忘记，在提倡体验不规则和自然性时，艾迪生在《旁观者》发表的小品文探讨了"想象的乐趣"，并对适用于野性自然和天然花园的美学形式进行了讨论。在 18 世纪上半叶，大家对于阿尔卑斯山的野外风景和景观花园内的自然风景有着相同的渴望。艾迪生辩称，在自然的广阔地域内，充满着各种类型的景色，旅行者可以自由地漫步。可以说，自然景观以及即将出现的画境园林景观给人带来的愉悦感是相通的。 但是，虽然阿尔卑斯山的无规则和多样性与 Kent、Brown、园林设计师以及其他设计师所创作的景观花园存在相似性，这其中也存在诸多问题。欧洲花园受瑞士的影响不仅存在于景观花园中，还存在于其他景观设计中。在呈现湍流和瀑布时，18 世纪的园林简直可以说是直接模仿了阿尔卑斯的景观。另一个非常重要的现象是这些园林模仿了在当时极为流行的瑞士特色，如：瑞士的桥梁或瑞士的农舍。通过模仿瑞士，英国乃至整个欧洲大陆的景观实际上成了他们所模仿对象的一部分。

图 21

Not only parks, but entire regions were thus 'Swissized' around 1800. Once he got to know Switzerland after a walking tour in 1777 and having translated William Coxe's *Travels in Switzerland*, the Frenchman Louis Ramond de Carbonnières 'discovered' the Pyrenees in 1786. Applying his Alpine patterns and expectations to this newly discovered region, he became the 'inventor' of the Pyrenees (see his book *Observations faites dans les Pyrénées, pour servir de suite à des observations sur les Alpes*, 1789). Similarly, the interest of the British in their own landscapes, especially in Wales, Scotland or regarding the Lake District is a result of their earlier Swissomania (and the cult of the classical Italian landscape) and the attention they developed for topography, sublime landscapes and picturesque objects.

在 1800 年，不仅是花园，整个地区都深深打上了瑞士烙印。1777 年法国人 Louis Ramond de Carbonnières 在徒步旅行中了解了瑞士并翻译了威廉 • 考克斯的《在瑞士旅行》后，于 1786 年发现了比利牛斯山。他把阿尔卑斯山的模型及对其的期望应用到这一新发现的地方，他成了比利牛斯山（参考其著作 Observations faites dans les Pyrénées, pour servir de suite à des observations sur les Alpes, 1789 年）的"发明家"。同样地，英国人对自己国家景观的兴趣，如对威尔士、苏格兰或湖区的兴趣，也源于最初对瑞士景观的兴趣，后来，他们开始注意本国的地形特征、崇高景观以及其他如画般的风景。

The 19th century, however, had less to do with changes in imagination and more with very real transformations in the Alpine regions of Switzerland (dialectically, they again

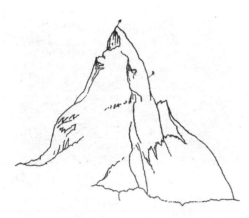

图 22

图 23

transformed the idea of Swiss Nature). One of the main factors of this change was tourism. At a time when huge hotels (ill. 22, ill. 23) started to occupy the highest regions of the country (Rigi, 1816; Faulhorn, 1823 – the highest hotel in Europe; Wengeralp, 1835; Kleine Scheidegg, 1838; Rothorn, 1840), the tourist industry was the main force setting the rules for gaining access to the sublime beauties. The hotel room, with its balcony and viewing points, became the privileged site where travelers could safely frame Nature. The possibility of being next to Nature became extremely easy, but it lacked the surprise of the earlier period. The arrival of the train (Vitznau-Rigi Kulm 1871) and the first collective trips to Switzerland (Thomas Cook 1858) brought more and more visitors into the remotest parts of the country and led to more standardization.

在19世纪，虽然相比而言想象力方面无太大变化，瑞士的阿尔卑斯山地区正经历着实实在在的转变（辩证地说，这又一次改变了自然状态下的瑞士景观）。促成这个变化的一个主要因素就是旅游业。当时，大型宾馆（图22，图23）开始占据该国最高的地区（瑞吉山，1816；福尔山，1823——欧洲地理位置最高的宾馆；Wengeralp，1835； Kleine Scheidegg，1838；洛特峰，1840），旅游业成为接近自然之美的主要途径。旅馆客房配备有观景阳台，旅行者可以在这里安全地欣赏自然景观。接近自然变得极其容易，但却缺乏早期自然带给人们的惊奇感。火车的通行（Vitznau-Rigi Kulm，1871）以及去往瑞士的首次集体性旅行（托马斯·库克1858）使越来越多的游客可以进入这个国家最遥远的区域，这也导致了旅游路线标准化的出现。

Another important factor at the time, with a deep impact on Nature, was the regulation of rivers and mountain torrents throughout the country. These measures enhanced the control of water, but the new canals and silenced rivers no longer looked like they used to. Between, let's say, the formidable Staubbach falls, described by every guide book and painted by generations of artists and the reality on the terrain the gap widened, and especially so with the hydro-electric schemes developed at the end of the century. During this period, almost every wild element was tamed and adapted to manmade design. The fact that tourism, water control, energy production and the transport systems conquered the totality of Nature raised some criticism, but generally it was accepted in the name of progress.

当时，对自然产生深远影响的另一个重要因素是遍及全国的河道整治和山洪治理。这些措施增强了对河水的控制，但是新开凿的运河和治理后的河流再也不像过去那样了。举例来说，令人生畏的施陶河瀑布，几乎每部旅行指南都描述过，一代又一代艺术家也都曾描绘过，随着世纪末的水电开发而失去了原来的样貌。在这个时期，几乎所有的野性元素都被驯化，被人为设计。旅游业、水利措施、发电以及运输系统的发展破坏了自然的整体性，并引发一些批评之声，但是总体而言，这还是以发展的名义而被接受。

Only solitary voices like that of John Ruskin (ill. 24, ill. 25), who toured the Alps for more than three decades, drawing, painting, photographing and writing about his beloved mountains, dared to attack the thorough and definitive transformation of Nature. Ruskin or the German inventor of Heimatschutz, Rudorff, advocated a comeback to pristine Nature, throwing a nostalgic light on it. One of the mythic and nostalgic ways to express the longing for a primeval Nature at the moment of its loss was the well-known *Heidi*. Johanna Spyri's tale (she adapted an earlier novel by a German writer, Adelaid: *the girl from the Alps*, 1830) of the wonderful world in the Grisons where Heidi and her grandfather – a sort of neohallerian incarnation of the good wild man of the mountains – live free of civilization in the midst of sublime scenery. *Heidi* was written by an author raised in urban Zurich and was imagined at a time when the city had already won the battle against Nature. The innocence of the goat milk-drinking girl, the effect of the fresh air on her friend Klara, the sense of pride, independence and liberty of her grandfather – all this became an international success story (ill. 26, ill. 27, ill. 28) and a national myth at a time when Nature was definitively controlled by technology, economics and the power of urban civilization. As in 14th century Siena or in 18th century Bern, with Haller, or in Geneva and Paris, with Rousseau, or again with Ruskin and Rudorff: the demand for Nature and the nostalgic projections on it always have their origins in the city and in its complicated relation to its 'other': to the countryside and to the deserts (for a long time the word desert meant voids).

图 24

图 25

只有少数的反对声音敢于攻击这种对自然的彻底、人为的改变。约翰 • 罗斯金（图24，图25）就是其中之一。30多年间，他在阿尔卑斯山脉游历、绘画、照相并写作。罗斯金、德国发明家 Heimatschutz 以及鲁道夫提倡恢复原始自然，颇具怀旧情怀。在失去时，以神话般怀旧的方式把对原始自然的渴望表现得最淋漓尽致的当属著名的《海蒂》。这是乔安娜 • 斯比瑞（她改编了由一位德国作家 Adelaid 所著的一部早期小说——《来自阿尔卑斯山的少女》，1830）所著的一部小说，描绘了海蒂和她的祖父在位于 Grisons 地区的奇妙世界里的生活，他们就像是野性自然的化身，生活在不受现代文明干扰的崇高风景中。《海蒂》的作者在苏黎世市区长大，当时城市已经赢得了与自然之间的斗争。这位天真无邪、喝羊奶长大的小姑娘与她的好朋友克拉拉为读者带来的清新之风，以及祖父的倔强、追求独立与自由的性格让这部小说登上全球长期畅销书榜（图26，图27，图28），也成就了一段瑞士神话。因为此书面世之时，瑞士的自然正完完全全地受制于科技、经济与城市化之下。就像14世纪的锡耶纳或18世纪的伯尔尼，或是日内瓦和巴黎，如哈勒、卢梭、罗斯金和鲁道夫所说的那样，对自然的需求以及对自然的怀旧情结往往起源于城市，并与城市相对的乡村和沙漠（长期以来这个词的意思是空无）有着复杂的关系。

What, then, is a Swiss landscape, or does such a formula make no sense? The first answer is certainly that there are no invariable Swiss landscapes, and that even the way in which the most solid and permanent elements of the Swiss territory, the mountains, are perceived, changes permanently. Landscape is indeed closer to the concept of difference than to the opposite, that of identity. Before 1700, when wild Nature was not yet an object of interest, almost the entire Alpine realm was out of sight. At that time, only tamed Nature, that is the gardenlike sites of the country downhill or a beautiful city could appear, and even these rarely, as something beautiful and interesting enough to be studied. In the 1750, there was already a sort of catalogue of places worth visiting because of their outstanding natural qualities. At that time, however, there were still huge differences between the landscapes of a minority (poets, painters, scientists, etc., people with a specialized 'way of seeing') and the masses, which were not yet interested in natural beauty. There is an important difference too between the attitude of curious foreigners and more conservative residents. During the second half of the 18th century, landscape became fashionable. Fortunate tourists knew how to identify beautiful, sublime and picturesque objects, and came to visually 'hunt' for them in Switzerland. In a period already characterized by the massive circulation of representations, travelers rambled around, compared, judged, described and drew.

那么，什么是瑞士景观，有没有标准答案？第一个答案是，当然没有不变的瑞士景观，即便是瑞士地域内最坚实和持久的元素，如山脉，也是处在永恒的变化中的。景观的确正在走向差异化，而不是相反的方向，即同一性。在1700年之前，野性自然还不是兴趣的目标，几乎整个阿尔卑斯山区域都不在人们的视线范围之内。那时，只有经过驯服的自然，也就是人工花园或美丽的城市才能引起人们的兴趣，吸引人们去研究它。1750年就已经有了专门介绍值得游玩的地方的目录，因为这些地区具备优良的自然品质。但是，在当时，少数人（如诗人、画家、科学家等，以及有独到见解的人）眼中的景观与大众所谓的景观仍存在巨大差异，因为大众还未对自然之美产生兴趣。好奇的外国人与相对保守的当地居民在态度上也存在着重要区别。在18世纪下半叶，景观变得流行起来。幸运的游客知道如何定义美丽的、崇高的和

图26

图27

如画的景观，他们来到瑞士，为了亲眼看看这些景观。在当时景观表现方法广泛流行的时期，旅行者四处观赏、对比、评价、描述并描绘景观。

Generally speaking, a significant part of the land becomes a landscape only because it is recognized as such. The interest for landscape is always mediated and motivated by other factors: one wishes to see the 'wonders' described in the guide books or sung by the poets, or one desires to share the 'same' landscape as Saint-Preux or other fictional characters. In other words, and even after having accepted Nature as positive, it took a long period and a real 'education' in the way of seeing the world in order to frame it and to constitute landscapes. The 'free' landscape as the result of surprise is something very recent and very rare.

图 28

一般而言，一个地方的某个地区能成为景观的原因只是因为它被认为就是那样的。对景观的兴趣往往是受到其他因素的介入和刺激，人们希望看见旅行指南中描述的或诗人赞颂的"奇观"，或者说，人们希望分享圣·布莱尤思或其他小说人物所经历的"相同"景观。换句话说，即使是已经接受了自然是积极的这一理念，仍要经历很长时间和受到看待世界的有效"教育"方式后，才能表达自然，并构造景观。由惊奇带来的"自由"景观是新兴事物且非常罕见。

图 29

Around 1800, most of the cultivated Europeans (and even many non-Europeans) had plenty of sublime and picturesque set images in their mind: they knew and enjoyed Nature both typologically – they could recognize a particularly savage torrent, an amazing precipice, an overwhelming site, etc. – and specifically: they all had an idea, even if they had never been there, of the wonderful Staubbach or the falls at Schaffhausen, the Jungfrau, the Rigi and of many other prominent places. Around 1900, the situation was already quite complex: while infrastructures, tourism and hydroelectricity had definitively transformed almost the entire territory of Switzerland into one gigantic machine (ill. 29), many people persisted in looking for pristine Nature even in places where every step reminded them of the human presence.

约在 1800 年，大部分受过教育的欧洲人（甚至是许多非欧洲人）在脑海中有许多崇高的、如画的想象，他们以类型学的方式了解并欣赏自然，他们可以识别独特的原始湍流、令人惊愕的悬崖以及令人折服的场景。具体地说，即使他们没有去过，对于惊奇的施陶河瀑布或沙夫豪森瀑布、圣母峰、瑞吉山以及许多其他地方，他们都有着自己的想象。约在 1900 年，情况已经变得相当复杂，公共设施、旅游业和水电站几乎完全改变了整个瑞士的地域面貌，把它变成了一个巨大的机器（图 29）。许多人坚持寻找原始状态的自然，然而所到之处无不提醒着他们人类干预的存在。

This contradictory situation lives on well into our days, when Bollywood filmmakers come to the Swiss Alps in search of pure mountains and wonderful scenery and when the global tourist industry brings millions of visitors to the country still seen as a sort of "wonderland".

这种矛盾在我们生活的当下仍然存在，当宝莱坞电影制作人来到瑞士阿尔卑斯山寻求"纯正"的山脉和惊奇的风景时，当全球旅游业把数百万游客带到瑞士来时，瑞士仍被认为是一个"奇境"。

The Swiss landscape – something dynamic and never simply given – is therefore the result of a process, where after a long period of construction (consolidating landscape patterns, teaching how to recognize sublime and picturesque features) comes deconstruction (a sense of repetition) and, after that, reconstruction once again. The "Disneylandization" of Switzerland, put forward by the Swiss sociologist Bernard Crettaz, is therefore nothing really new. The ideological production and re-production of the image of the country existed from the beginning, probably as an effect of the bourgeois bad conscience, when confronted with the loss of Nature in a world dominated by technology and commerce.

所以，瑞士景观———一种动态的且从来不是简单赋予的——是过程的结果，经过长时期的构造设计（整合景观格局，教授如何认识崇高和如画般的景观特色），带来的是再次解构，然后再一次地重建。如瑞士社会学家Bernard Crettaz所提出的那样，瑞士的"迪斯尼乐园化"根本毫无新意可言。一个国家的形象从最初意识形态不断斗争中就已存在，在世界受到科技和商业的统治时，自然面临丧失，这或许是商人居心不良的结果所致。

Should we therefore speak of Swiss landscape architecture or of landscape architecture in Switzerland? There is no simple way to answer this question, and this is not only linked to the particularities of Switzerland, but to the specificity of landscape architecture as such. The discipline of landscape architecture is indeed – something we should always bear in mind – a very recent one. It started shortly before 1800 in England (with Humphry Repton) and in France (with Jean-Marie Morel). However, it never developed continuously in these two pioneering countries. Generally speaking, landscape architecture is marked by temporal and regional discontinuities and by its fragile identity. This almost permanent state of crisis and its ongoing struggle with other disciplines (above all, with architecture) is, nevertheless, at the same time its strength. The possibility to work on the edge, to take care of spaces in between sites, buildings, places, disciplines and practices, has provided landscape architecture with disciplinary freedom and a sense of openness. Staying at the margins and crossing disciplines implied, on the other hand, a lack of visibility, something quite surprising at first sight if we think how thoroughly the world has been landscaped during the last two centuries, whether by landscape architects, architects, engineers or other protagonists. Even the history of the discipline has still to be written and there are more open questions than answers concerning its field, its methods and its relation to other disciplines.

那么我们是不是可以说景观设计学是瑞士所特有的，或者说景观设计学很适合瑞士？对于这个问题，没有简单的答案，这不仅关系到瑞士的特性，还关系到景观设计的特殊性。景观设计学（这是我们应该牢记的）的确是近代兴起的学科。约1800年，它萌发于英国(汉弗莱 • 雷普顿)和法国(让－玛丽 • 莫雷尔)。但是在这两个起源国家，它的发展并不是连续的。一般而言，景观设计带有区域和时间的不连续性和脆弱的特性。它几乎处于永久的危机状态，且时刻在与其他学科斗争（首先是建筑学），但是同时，这也是它发展的动力。站在学科前沿以及不断处理场地、建筑、区域、学科和实践之间的关系赋予景观设计学充分的自由和包容性。 处在学科边缘并与其他学科交叉，从另一方面来说，看似缺乏清晰度，但是考虑到世界已被景观彻底改变这一事实，我们不得不对景观设计学感到惊讶。的确，在过去的两个世纪，这种改变不仅是由建筑师和工程师实现的，还有景观设计师。然而这一学科的历史还有待书写，关于该学科的领域范围、方法以及与其他学科的关系等问题还有很多需要解决。

In Switzerland, landscape architecture is even more recent: in a radical sense, it all started only after World War II. This reduced time span is, however, indirectly proportional to the achievements of Swiss landscape architecture: a few decades have permitted the rise of an extremely mature profession with important personalities and fundamental projects having a significant impact on the international scene.

在瑞士，景观设计兴起得也比较晚，可以说，它是在第二次世界大战后兴起的。虽然时间较短，但是瑞士景观设计学所取得的成就确实是巨大的。仅仅几十年就诞生了一批职业化的、著名的园林景观设计家，他们给国际景观设计带来巨大影响。

While such a short timeframe (1945-today) may seem plausible and supported by facts, things appear to be more complicated when we take a closer look into the matter. First of all, and as we have shown before, Switzerland was already present outside Switzerland as a set of patterns and as an aesthetic concept: the idea of Switzerland

图 30

图 31

did not only influence the rise of tourism, but it functioned as an international design scheme as well (Swiss bridges, Swiss chalets, Swiss waterfalls, etc. present in picturesque gardens throughout Europe). There exists, furthermore, an important prehistory of Swiss landscape architecture, linked especially to exhibitions. (ill. 30, ill. 31, ill. 32) The Swiss National exhibitions and the Garden Exhibitions contributed throughout the 20th century to the transformation of the activity of modern "gardeners" into landscape architecture. Furthermore, landscape architecture hasn't only developed within the discipline itself, but also next to it, in related fields, for example in architecture (one has to think only of Le Corbusier's interest in landscape and even in landscape architecture).

如此短的时间（从 1945 年至今）就能取得这样的成就看起来似乎是合理且有事实支撑的，但如果我们仔细分析一下，事情可能变得复杂得多。 首先，如我们前面所述，在瑞士以外，瑞士已经被认为是一种模式和美学概念，瑞士理念不仅造就了旅游业的兴起，它还被作为一种国际化的设计方案（瑞士桥梁、瑞士农舍、瑞士瀑布等在整个欧洲的花园设计中被模仿）。其次，瑞士景观设计学有着重要的历史背景，尤其与各种展览关系密切（图 30，图 31，图 32）。在整个 20 世纪瑞士国家展览会和园艺展览会在把现代"花园"改变为景观设计的过程中起到重要的作用。另外，景观设计学不是孤立发展的学科，它与其他学科，如关系最近的建筑学，有着千丝万缕的联系（只要想想建筑师勒 • 柯布西耶对景观和景观设计的浓厚兴趣就可看出这一点）。

To work today as a landscape architect in Switzerland means to be part of a complex historical palimpsest. The Alps for instance – a permanent presence and the horizon of Swiss self-consciousness – are both the ultimate monuments of Nature and its opposite, that is faux mountains: they are indeed manmade objects as well. Swiss Landscape architecture is, therefore, deeply rooted in the geological, historical and orographic tradition of its territory, whilst also being international, eclectic and open minded.

在瑞士作为景观设计师意味着要成为复杂的历史重写本的一部分。例如阿尔卑斯山脉，是一个永久的存在以及瑞士人自我意识的地平线，它既是自然最后的纪念碑也是自然的对立面，也就是说人造山，它们也的确是人造对象。所以，瑞士景观设计深深扎根于其地域的地质、历史和地形传统中，但同时也是国际化的、折中的和开放的。

However, despite its great variety, most of the projects imagined and built during recent decades share at least three qualities or values: rigor, elegance and ecological consciousness.

尽管存在很大的差异性，最近几十年来景观设计项目却共同具备三个品质或价值：严格、优雅及生态意识。

There is a sense of moderation and strictness, which dominates the works of Swiss landscape architects (ill. 33, ill. 34, ill. 35) no matter where or how they plan. Theirs is a touch of minimalism, an ascetic and almost 'protestant' attitude. They privilege simple and expressive materials and even the most complicated of their projects appears as the result of necessity. Not to put forward the individual persona and signature, but let the site, the materials and the form speak, shows that despite all of the discontinuities of history, the idea of necessity (Haller thought of such an attitude as typical of the Alpine dwellers), of creating strong effects by using as few elements as possible, has survived even in the postmodern world.

图 32

在瑞士景观设计师的工作中，无论在哪里设计或是怎样设计，适度和严谨意识都占据主导地位（图 33，图 34，图 35），还有些许极简派、苦行僧式以及近乎"新教徒式"态度。他们热衷简单且表现力丰富的材料，甚至他们设计的一些复杂项目能够借此呈现出来。他们不注重个人印记，而是让场地、原料和形态自己"发言"，宣示着历史上曾中断的，通过使用最少的元素打造出强大效果的这种思想（哈勒认为的阿尔卑斯居民的那种态度）仍存活于现在的世界中。

图 33

Another fundamental characteristic of Swiss landscape architecture is its struggle for the best formal solutions. The modern idea of "good form", linked to the Werkbund and the Bauhaus, is very much present in the elegant solutions of most projects in recent decades. The best way to define this particular form or Gestalt of the individual works is to say what they are not: they are not overloaded (with material, sense) and they never appear as the consequence of instant, rapid solutions. Their "good form" is the result of a practice that takes time, a lot of time, of a creative process where the slow maturation of the design process is more important than exuberant intuition.

图 34

瑞士景观设计的另一个基本特征是竭力寻求最佳形式的解决途径。"最佳形式"这种现代理念，与德国工业同盟的现代主义设计和包豪斯风格有一定的关联，最近几十年在大部分项目的解决途径中都广泛存在。定义这一最好形式或某一项目形式的最佳方法是说明它们不是什么，如它们不是过度设计（在材料和感官上），它们不是匆忙草率中想出的解决途径。这种"最佳形式"是长时间实践的结果，是创作的过程，慢慢成熟的设计过程要比凭直觉设计重要得多。

图 35

Swiss landscape architecture is, naturally, not a priori ecological. The presence of Nature, its function both as a set of images and as a general concept, the idea of Switzerland as the 'garden of Nature' and the permanent building and rebuilding of Nature represent a common ground and a point of departure for everyone working in Switzerland and shaping its landscapes. Swiss landscape architecture appears to be ecological not by simply applying abstract ideas such as sustainability, but rather by relating every singular project to the global idea of Nature, and specifically, to the temporality and metamorphic qualities of Nature.

瑞士景观设计是自然的，但并不是先天生态的。自然作为一系列影像和一般概念而存在，瑞士作为"自然花园"的这种观念，以及对自然的永久性构建和重建代表着在瑞士工作和塑造景观的每个人的共同点和出发点。瑞士景观设计的生态性并不是简单地通过利用一些抽象的概念表现出来，如可持续性，而是把每个项目都与大自然的全球理念联系在一起，具体地说，就是与自然的时间性和变化性特质联系在一起。

Many important projects realized during the last fifty years have taken into consideration complex time layers and processes (ill. 36, ill. 37) . By doing so, these projects haven't become 'natural' and they haven't applied for an ecological label either; they have rather appeared as symbols of the ecology of the future or, to put it another way: they are living metaphors of life.

在过去 50 年间所完成的许多重要项目都把复杂的时间层和过程考虑在内（图 36，图 37）。这样做，这些项目不是为了标榜"天然"，也没刻意地使用生态这一称号，它们更多地表现为未来生态学的象征，或者换句话说，它们是生命活生生的比喻。

图 36

参考文献

[1] J. Dennis. Miscellanies in Verse and Prose (1693). The Critical Works of John Dennis [M]. Baltimore: The Johns Hopkins Press, 1943.

[2] E. Burke. A Philosophical Enquiry into the Origin of Our Ideas of the Sublime and Beautiful [M]. Oxford: Oxford University Press. 1757.

[3] Edward Hamley. Translation from the Alps of Haller [M]. Poems of Various Kinds 1795.

[4] Joseph Addison. The Pleasures of the Imagination [J]. The Spectator, 1712.

图 37

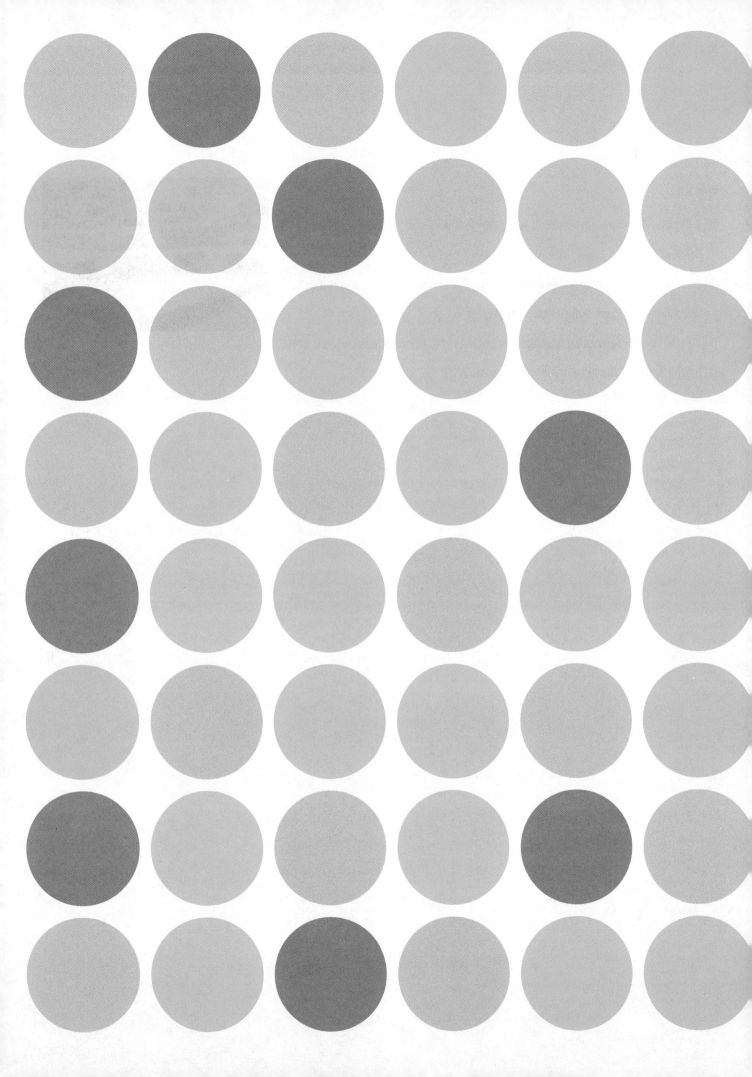

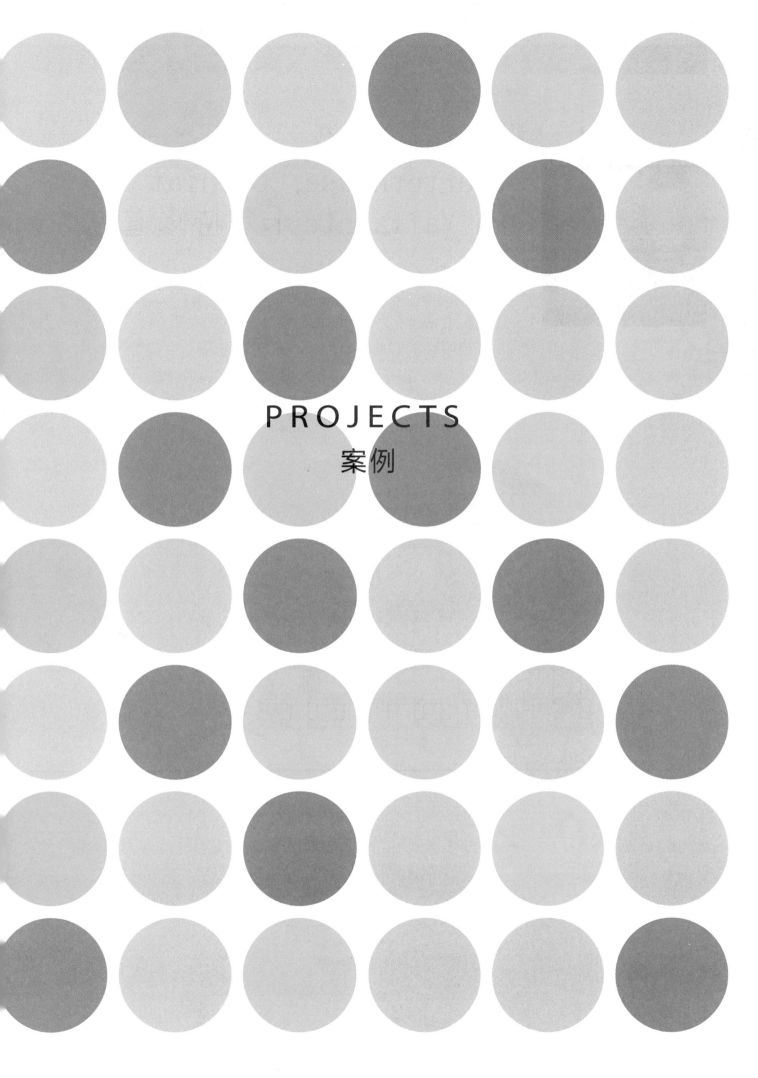

PROJECTS
案例

La Valsainte Charterhouse, Cerniat
La Valsainte 卡尔特修道院

项目地点：Cerniat

设计单位：Hüsler & Associés and Pascal Amphoux Contrepoint, Lausanne

项目面积：10 000平方米

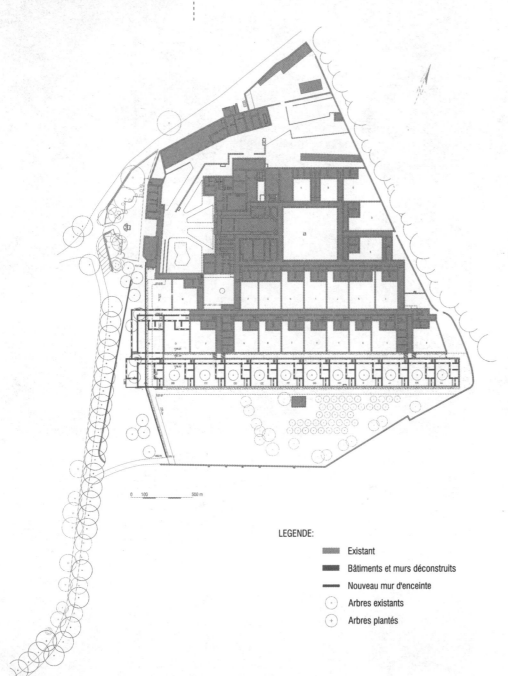

LEGENDE:
- Existant
- Bâtiments et murs déconstruits
- Nouveau mur d'enceinte
- Arbres existants
- Arbres plantés

经过工程师们的技术分析，确认需尽快清理地下室，另外，为了最大幅度地减少开支，需拆除14个修道士居所，这些居所修建于20世纪初，位于修道院的南部前端。这项工程摒弃了过度重构，重点突出了该修道院的历史。整个建筑由三个子空间构成。Terrasse是一个巨大的内部空间，两边只是简单地种植了果树，令人们回想起被拆除的14个修道士居所。每棵果树代表了死后的生活，位于每个花坛的正中央。Talwegis是过渡空间，植有枫树丛，它延长了进入修道院的道路。拆除修道士居所后产生的废料用于修建外墙，并于19世纪末进行了改造，使其极具现代风格。大道最终重新被绿荫覆盖，成为一条植物拱廊：对椴树大道进行了扩展和延长。当到达La Valsainte时，就会产生一种新的时间观念，因为它提醒驾驶者减速，欢迎步行进入，并为参观者展示出一幅幅修道院的真实图。

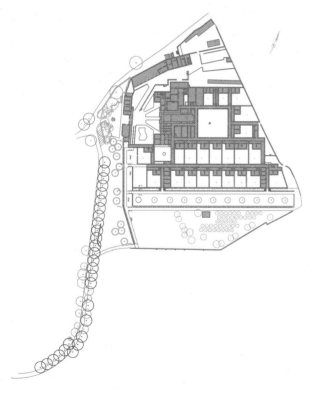

VEGETATION

- Alignement, erables et tilleuils
- Les bosquets, érables et frênes
- Les fruitiers

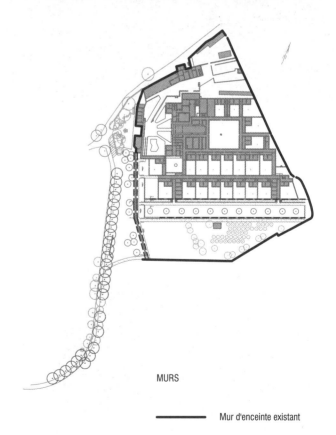

MURS

- Mur d'enceinte existant
- Mur d'enceinte reconstruit (sur le tracé de 1865)
- Mur de bâtiment converti en mur extérieur

Rudolf-Bednar Park, Vienna
维也纳鲁道夫-贝德纳公园

项目地点：维也纳

设计单位：Hager Partner AG, Zurich

项目面积：32 000平方米

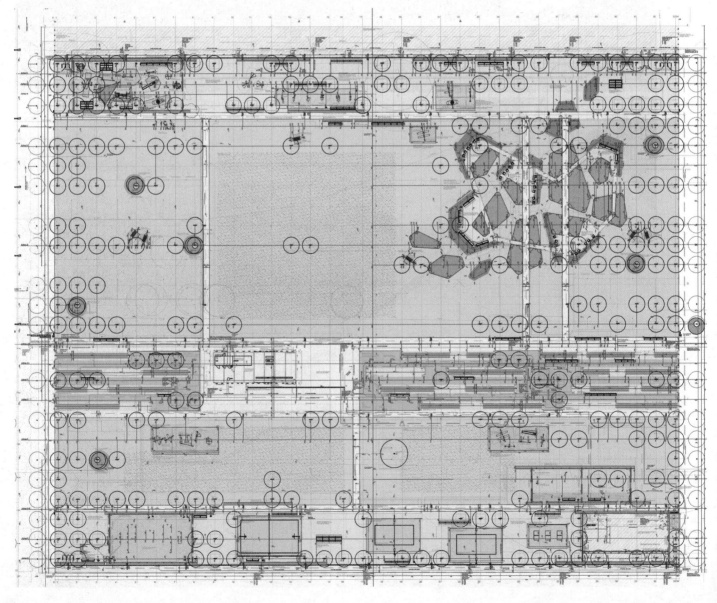

在维也纳的利奥波德城铁路区修建新公园是整个区域改造的起点。象征活力和传统的菩提树形成了一段具有纪念意义的道路，其与已弃用的铁轨的长度和方向相一致。整个场地给人紧凑的感觉，和谐地将公园与整个街区融为一体。一棵棵白杨、皂荚、红枫和针栎营造出充满浓厚回忆色彩的氛围。它们形成了一方植物帷幕，环绕着操场、运动场、儿童区和操场旁边的座椅区。橘红色的木桩指向活动更加激烈的区域，一棵棵树聚在一起，形成了一个个小树林——其中点缀着一些林间空地。在公园中央，芦苇园与多瑙河支流形成的原始风景交相呼应。

043

| THE SWISS TOUCH IN LANDSCAPE ARCHITECTURE | 景观设计中的瑞士印迹

Eulach Park, Winterthur
温特图尔 Eulach 公园

项目地点：温特图尔
设计单位：koepflipartner landschaftsarchitekten, Lucerne
施工时间：2007—2011
项目面积：70 000 平方米

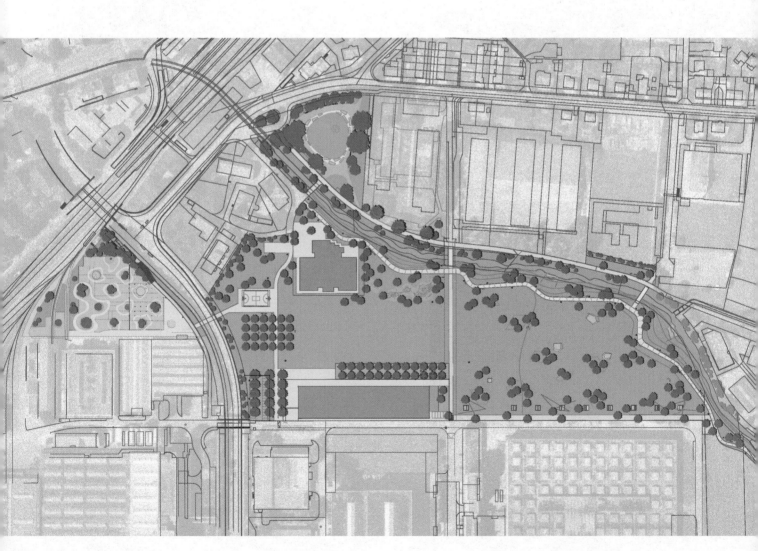

 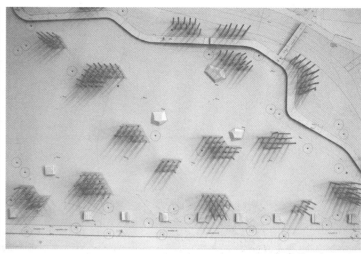

温特图尔有一片海拔较高的地区，那里曾经是苏尔寿工业用地，就在这片土地上兴建一个集住宅、服务、商业为一体的整改工程。Eulach公园是整个城市发展的中心和起点，这个公园代表着温特图尔未来的形象，带着浓郁的城市特色。

该公园占地70 000平方米，没有按照任何明确的几何形状进行规划。公园北部与新建的Eulach区毗邻，其他方向则与不同的街道和建筑物接壤。

公园最大的特点是其像一个植物园，一方面树木呈几何形排列，树冠相互交错，形成一片片巨大的树荫，另一方面在院内无序地种植了橡树。树木与雕塑作品和立体的空间让各个区域联系成一个整体。访客在不同的氛围和不同的空间密度中行走，总是能看到不同的画面。橡树是这个变化的景观的起点，橡树稀少的地方形成一幅稀疏草原的画面，而橡树密集的地方又呈现出另一幅画面，通过橡树稀密程度的不同为游客展现了一个动态的画面。

整个公园装饰相对含蓄。辽阔的草地、河流与公园采用天然石和混凝土做的防洪坡联系在一起，巧妙地利用混凝土、木材和草地为游客提供休息区，游客可在此休闲娱乐。

053

THE SWISS TOUCH IN LANDSCAPE ARCHITECTURE
景观设计中的瑞士印迹

The Tree Museum, Rapperswil-Jona
拉珀斯维尔-乔纳树博物馆

■ 项目地点：苏黎世
■ 主设计师：Enea GmbH, Rapperswil-Jona
■ 项目面积：75 000平方米

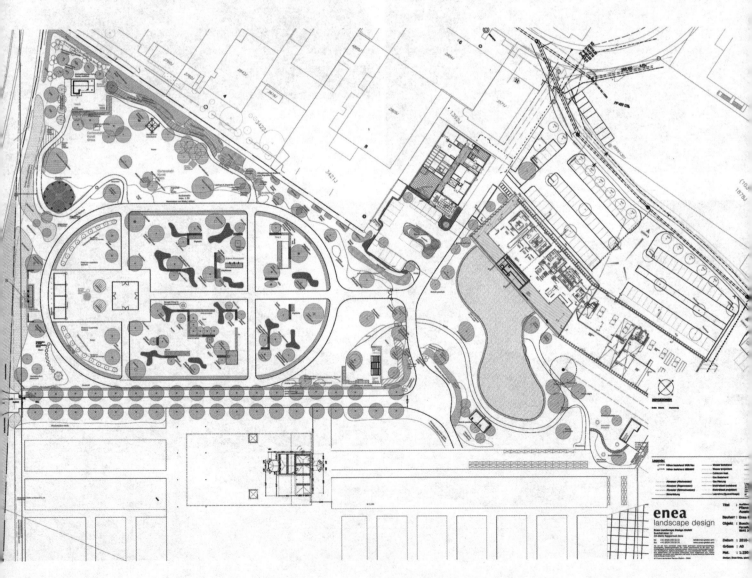

树博物馆被构思为一个椭圆形的露天博物馆，划分为一个个小空间，每个空间都具有自己的氛围和特点，树种来自Enzo Enea的收藏，Enzo Enea是一名瑞士景观设计师和著名的树种收藏者。博物馆的第一使命就是珍藏珍稀树种，展示出树种的美丽和稀有性。接下来，在更深层次上，Enzo Enea的收藏会帮助参观者形成对于生命的原始属性的认知，比如时间和空间观念，以及这些属性是以何种方式内含在这些古老、珍贵树种的元素里的。

建立一个树博物馆的想法是Enzo Enea作为景观设计师，其作品自然而然的延伸，他多年来通过对各种树的仔细观察和研究，将感知和照顾树木结合在一起，这不仅使他在这个领域声名鹊起，也使他对这些最平凡的自然造物充满了无尽的欣赏和敬佩。

为了将这些体验与更多人分享，Enzo Enea决定为他收藏的树修建一个博物馆，以表明他的收藏值得在博物馆里受到应有的关怀和关注，正如其他博物馆里的藏品一样。他修建露天"空间"的理念正体现了所有Enzo Enea花园的特点——允许树成为一个"个体"，参观者沿着这些空间，从不同角度去观察它们。

参观树博物馆带来的体验和感动来自于不同元素，包括树木自身的壮观、树木周围形成的小气候、纹理的多样性、间隔和比例的效果等。当然，大多数树木最为非凡和最令人感动的一个特点就是它们的年龄。"慢生活"需求的苏醒，对自然和环境的敬佩和欣赏，这些都是参观树博物馆得到的主要启发。其精神、地方风情，将会使我们潜意识里这些古老形状投射产生的意识"外化"。

树博物馆的珍品将近有50棵树，代表了25个树种，树龄超过100年的有好几棵。不朽的光环与"时间"意识，在我们忙碌的世界里弥足珍贵，而在树博物馆里却无处不在，这是一个令人沉思的净地。受古代盆景成型艺术影响的复杂技术被用来移植和保护树木。另外，在公园里环绕博物馆栽种了100棵其他树木和一些植物，充当景观设计和空间实验室。

博物馆和公园地区的独有树种加起来超过了2000种，是Enzo Enea在过去17年里收集的成果。他的收藏只包括属于该地气候带的树种。

公园的主要特征体现在2500平方米的Enea景观设计总部，在总部前面有一个杂草蔓生的火山岩层湖。这个建筑由美国建筑公司Oppenheim Architecture & Design建造，里面有花园设备精品展、一家博物馆商店，以及许多艺术和设计作品。这个建筑物被芝加哥雅典娜博物馆在2009年授予美国建筑大奖。

Enea景观建筑公司以前位于邻镇Schmerikon。后来拉波斯维尔的所有者Cistercian Cloister Wurmsbach愿意将土地租给Enea景观建筑公司，因此，公司才搬到了现在的拉珀斯维尔－乔纳。

Prime Tower, Zurich
苏黎世泛利大厦

■ 项目地点：苏黎世
  设计公司：Studio Vulkan Landschaftsarchitektur, Zurich
  建筑师：Gigon Guyer
  摄影：Thies Wachter, Daniela Valentini

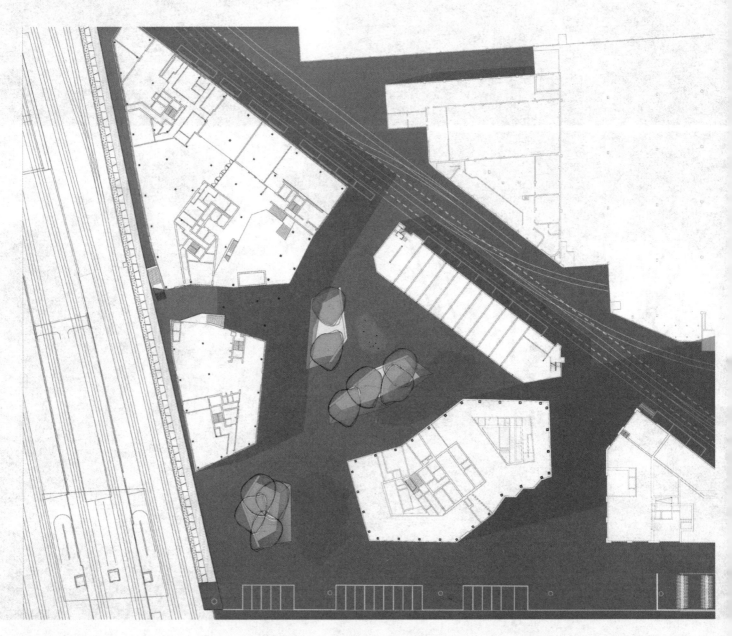

泛利大厦是一座外表面用玻璃幕墙装饰的高层建筑,主宰着苏黎世的天际线。现在,鉴于周围环境和实际需求,这座高层建筑的周围空间被改造成了一个流线形的广场,与泛利大厦和其他建筑群之间开阔的庭院实现了无缝对接。

围绕着建筑群的柏油路,形成了苏黎世的城市地表,体现了城市规划的基本原则,也是泛利大厦周围户外空间的指南。在庭院里,柏油路地表形成三个波浪,并被分割成一块块小型的绿色空间,绿草树木从凹陷的花坛中长出。地表上随意种植的白杨树显得十分柔弱,通过大厦表面玻璃幕墙的反射,显得比比皆是。从外面看,有着弧度的柏油路表面只是一个个斜坡,而从里面看,宽阔的水泥墙成了它的框架。在透明和高度功能化的城市空间之中,处处考虑了人体尺度的细节。每个柏油坪都有绿草、绿荫树、座椅和水元素,吸引人们来到这片空间内。

汽车、出租车和自行车停车场的设计都最大限度地考虑了透明性和渗透性。进入停车场的路径由可收回护柱控制。实用需求和柏油岛背后的城市基本原则营造出一个便捷却维持着低交通流量的空间,这个空间犹如建筑结构与人类之间的中介。

Market Place, Renens
勒南市场

项目地点：勒南
设计单位：Paysagestion, Lausanne
项目面积：2700平方米

勒南，前工业小镇，经过几年的改造，现在已成为城镇的一道风景线。瑞士文化遗产组织认可其开发的独特性，于2011年向其颁发了华克奖。位于项目中心处的市场，是与居民一起协调完成的，融入了双方的心血。市场是一处朴素、简单的空间，为日常生活及特殊事件提供场所，将乡村化的过去与城市化的现在紧密地联系在了一起。它唤起人们对犁地的记忆，在犁过的土地上，"禾苗"破土而出：市场交易、音乐、集会、演出及比赛。市场的地下是泥土及树木的根系和地下停车设施。地上，铺路石嵌在双色水泥里，高高的屋顶，一些长凳及一个舞台，重复的纵向形状仿照了以前耕地的沟垄。一块净地，可散步，跳舞，闲坐，观赏，野炊，用来沉思放松，看白云飘过。一块边界明显的净地，一些地方由粗糙的材料修成，易于维护。这是附着在城市风景中的一块田园净地。

Bear Park, Bern
伯尔尼熊公园

项目地点：伯尔尼

景观建筑师、建筑指导：Klötzli Friedli Landschaftsarchitekten AG, Bern

建筑设计：Matti Ragaz Hitz, Bern

工程设计：WAM Partner, Bern

艺术设计：Sibylla Walpen, Bern

摄影：Architekturfotografie Gempeler and Matti Ragaz Hitz Architekten AG

介绍

在伯尔尼市,在市中心养熊有着悠久的历史。现在的熊坑是一个纪念馆,对瑞士来说具有重要的意义,其是在1857年开放的。算起来它其实是第四处熊坑地点。

16世纪的记录显示第一处熊坑位于监狱塔要塞的前方,今天,那里成了"熊广场"。

熊坑是伯尔尼市具有国际知名度的旅游胜地。伯尔尼市当局认可了游客们提出的在熊坑中人道化养熊的意见。于是,伯尔尼市决定扩大熊坑以满足现代旅游业的需求,并于2003年宣布进行新熊坑设计国际大奖赛。

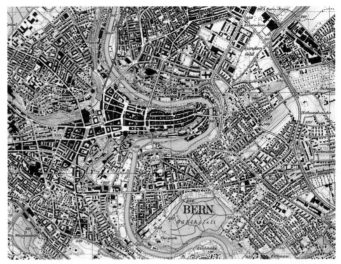

整体概念

阿勒河斜坡是市区的敏感地带，在此处，景观和城市和谐共融。

为了维持这种平衡，需要以紧凑型的方案来设计熊围场，需要围场与周围环境以及斜坡、河堤上的建筑物融为一体。

保持以前游客能近距离接触熊的便利，以及注重熊坑的保留。

操作

熊围场，占地6000平方米，准备放养一对成年熊和一到三只小熊。围场被分为两部分，以便将成年公熊和母熊及熊崽分开。熊全天都待在围场里，熊围场日夜开放。围场有导游服务，为游客讲解关于熊以及伯尔尼市的逸闻趣事，在围场也可以用手机下载相关资料。

围场有三个特别准备的熊洞，装有监控摄像头，为冬眠期和产崽期的熊提供住所。游客们可以通过监视器观看上述场景。

艺术

沿着熊公园的滨河小道装有玻璃护墙,可以使人们尽可能清晰地观察熊的洗浴区和公园其他的地方。为了保护小鸟不会撞上玻璃护墙,玻璃护墙不能是全透明的,所以设计师们在玻璃护墙上涂上了黑白玻璃涂料,带有伪装网图案。这种设计在环境和植被之间营造出一种有趣的视觉效果,与熊公园和谐共存。在公园里,熊能够隐藏自己,就如在自然环境中一样,在其中,游客们的角色变成了旁观者或是隐藏的观赏者,实现了公园的主题思想。伪装主题也用在了周围的水泥墙上,采取了精美的雕刻形式,使得整个场景和谐完美。

THE SWISS TOUCH IN LANDSCAPE ARCHITECTURE 景观设计中的瑞士印迹

Letten Swimming Area, Zurich
苏黎世 Letten 游泳区

项目地点：苏黎世

设计单位：Rotzler Krebs Partner Landschaftsarchitekten, Winterthur

项目面积：15 000 平方米

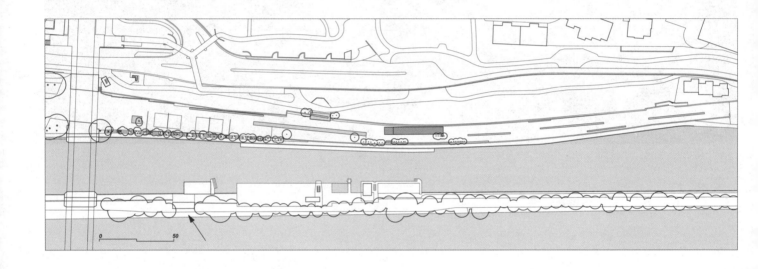

前铁路区的历史变迁，使得沿着供水渠的Letten地区成了当地居民、蜥蜴和杂草的家园。此项工程的目标就是将河流的集中娱乐用途与该区域作为大量蜥蜴集中栖息地的重要性结合起来。这些不同的用户有一个共同点——都喜欢太阳和温暖。短粗的铁道道砟为蜥蜴提供了生活的必要元素。在浴场铺上了光滑的沙路。两个区域之间没有设置墙和栅栏，以鼓励两边的居民自由穿越边界，如果天气不错，这种情况确实也会发生。为了强调区域的空间性，用于充当座椅的台阶由节奏性很强的加长楼梯元素构成，楔入现有树木之间，密集种植的桦树整齐地排列。这些元素相互贯穿，使人想起前铁路轨道上行驶着的火车。

Ruggächern Residential Complex, Zurich
Ruggächern 商住综合楼

项目地点：苏黎世

设计单位：Blau und Gelb Landschaftsarchitekten, Rapperswil

建筑师：Baumschlager Eberle

艺术设计：Markus Weiss

项目面积：28 200 平方米

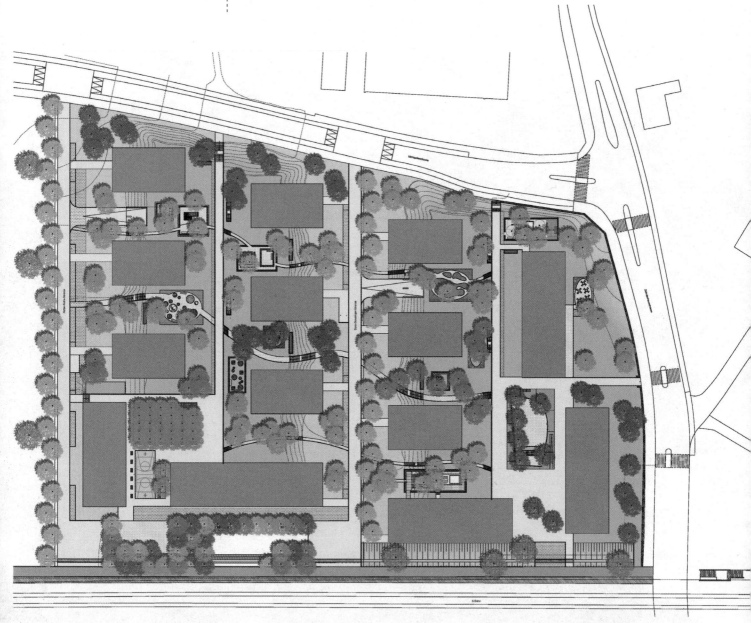

Ruggächern，这个大型商住综合楼坐落于阿佛恩德的郊区。阿佛恩德位于苏黎世这座国际化大都市的北部边界。阿佛恩德是苏黎世三大城市开发区之一——在过去7年里，修建了1300座新公寓楼，并计划在未来修建2100座新公寓楼。

Ruggächern商住综合楼周围由创造性的景观花园包围着，使得其中的人们犹如进入了一个迷人的公园中。Ruggächern商住综合楼包括5排建筑，为里面的9座塔式大厦构建了一个宽敞、宁静的空间。

建筑物入口有直行路、人行道和拱廊，构成了一个路径网络。公园中绿色空间的起伏使人犹如徜徉在公园之中；流动蜿蜒的人行小道一直伴你前行。

沿着蜿蜒的小道，你会被10座篱笆花园"小屋"所吸引。另外一个独特的地方就是这些花园都有着自己的主题：有的是一个运动场，有的是一个攀登园，有的是一个阅读园，甚至还有一个篱笆迷宫，一切都会让你流连忘返。

更为重要的是，每一个花园周围都有一个小树林与庭院相连，这使得平淡无奇的景观变得超凡脱俗。

这些花园里的植物都经过了精挑细选，以确保每个花园里的树木和植物同属一个种类，只不过形式不同。在某一个花园里你可能会发现欧洲角树和篱笆植物；而在另外一个花园里，你可能会发现酸橙树和篱笆植物。

为了提供顶级的自然体验,部分篱笆为双层篱笆,外层为连香木,内层为欧洲角树。通过这种方式,迷人花园的外观由内而外发生了变化。这确实是一种特殊的方式,带给人们最完整的自然体验,同时传递"住在花园里"的认同感。不言而喻,这种景观体验强化了温暖、棕红的秋季色彩和明亮、令人振奋的春季色彩。

向南走过去,你会发现更多可供消遣的地方——两个广场。西广场有一个舞台一样的空间,有着高高的酸橙树屋顶,还有一个球场,可以踢足球、打篮球。东广场,你会发现一个法式滚球场,栽有法式梧桐,还有一个供儿童玩耍的小型操场。迷人的绿色花园,到处充满着惊喜,Ruggächern无疑会让你魂牵梦绕,久久不能忘记。

THE SWISS TOUCH IN LANDSCAPE ARCHITECTURE | 景观设计中的瑞士印迹

Spreebogen Park, Berlin
柏林施普雷河湾公园

项目地点：柏林

设计单位：w+s Landschaftsarchitekten AG, Solothurn
　　　　　（ehemals Weber und Saurer）

客户：DSK Büro Berlin, Deutsche Stadt- u. Entwicklungsgesellschaft GmbH

合作单位：Gunter Frentzel, Raumkünstler, CH-Rüttenen Karin Lischner,
　　　　　Raumplanerin, Zurich

设计 / 施工管理：Gruppe F, Berlin

工程规划：WGG Schnetzer Puska, Basel und Leonhardt, Andrä und Partner, Berlin

项目面积：约 60 000 平方米

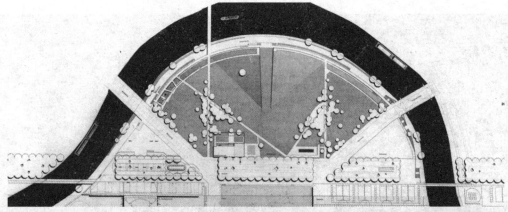

政府部门的新建筑标志着这个城市崭新的开始,赋予这个城市独特的"身份卡"。顺理成章,施普雷河湾公园的设计布局成为中心的议题。施普雷河湾公园历史悠久且不乏现代特色,它身上刻有年代的痕迹。基于公园的历史背景,公园的设计采用了新的园林建筑形式和艺术表现形式。

人工雕琢和自然美运用联系与对立、传统与现代、平凡与特殊这几种形式成就了一段佳话。过去、现在和未来之间的富有张力的关系成为整个公园设计的艺术理念:运用少数却极富表现力的元素来表现粗犷和内秀的特点。

水平和垂直的不同布局一方面很好地展现了公园的地理特征,另一方面也展现了其厚重的历史感及现代感。

施普雷河湾公园内部的树林、小路、草丛和凉亭融为一体，使得公园可以实现多种功能，比如有游乐运动区、音乐会表演区、日光浴区、放松休闲区和社交聚会区等。

新设置的叠嶂区位于公园的中部，因为中部相对于北部而言地势略微高耸，旁边的低洼地带刚好起到了"观景窗"的作用，两者共同打造出公园的新格局。两块钢板将这两个区域连接起来，从而与河岸和河港形成了一种奇妙的视觉效果，这样的设计可以方便人们穿行。与施普雷河之间明显的边缘设计再次使城市的北部边界独具特色。

在西部地势下沉的地带有一个花园（spurengarten）。三面均有桥梁可以通往此处，因而这个区域可以被视为一个通往过去的时间视窗：它包含了一个历史悠久的花园的一小部分。

在古老的天然石护坡上开辟了一个新的通道，可以让人们从花园走到河滨步道。雕琢精巧的石坡如同一座雕塑切断了与施普雷河的联系，彰显了新元素的力量。

Riedpark, Zug
楚格州里德公园

项目地点：楚格州

设计单位：Fontana Landschaftsarchitektur, Basel

洛尔桑地区的前身是瑞士经济增长最快的聚居区之一。这里有一个新的住宅发展项目"里德马特",这个古老走廊的名字诉说着充满水元素的自然景观的历史。整个设计项目围绕这个名字出发,希望达到与该地古老文化相契合的目的。附近的河流洛尔桑河为该地区的空间设置奠定了基调——一条蜿蜒的建筑带,将室外的空间划分成两个区域:市区、住宅区、街道位于北边,公园位于南边。该地区典型的湿地草原为公园的设计提供了范本。地下水距离地表仅有一米,所有规划必须精确考虑到这一特点,但这也为可持续性设计提供了契机。住宅街道区是住宿区、会友区、娱乐区、运动区的结合。街道两侧种植木兰科植物。地下车库的设计相对集中,这样的话楼房的对面就可以利用自然环境建立一所公园了。

芦苇属类草类和桤木群是湿地草原的典型植被,它们让公园再现了里德地区的原始风貌。里德公园里有一片不可通行的区域,里面生长着高达120厘米的芒草。这片高高的草丛让这片区域显得格外深邃,并且增强了地形建模的效果,给游戏休闲区划出了一个相对独立的场所。与芒草丛类似,公园中分散的水域也勾勒出一幅极具自然美的画面:地表圆形下凹的混凝土区可以收集地表水,有利于保持该地区地下水水位。这个公园内的经典设计元素和盆地一同起到了调解自然和公园地下水的作用。

Tessinerplatz, Zurich
苏黎世提契诺州广场

项目地点：苏黎世

设计单位：KuhnLandschaftsarchitekten, Zurich

客户：Tiefbauamt der Stadt Zurich

灯光设计：Priska Meier Lichtkonzepte Turgi

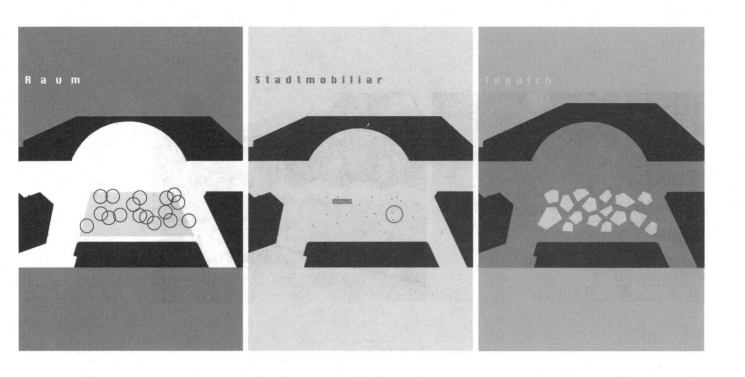

对于许多游客和外国打工者来说恩格火车站开启了一扇"一路向南的大门",从这儿他们乘坐火车返回家乡或者去往度假目的地。与此相反,归来的人则大都是从南部来的。作为一个主要发送远程客线的火车站,恩格火车站已经成为该地区一个非常重要的交通枢纽,日均乘客数量已达45 000人次。

提契诺州广场的重建应该是整合城市空间布局的第一个也是最重要的措施。恩格火车站始建于1926/1927年,由菲斯特兄弟设计建造。火车站前有一个半月形隆起的设计,与海洋大街相连,方便旅客通过。直到20世纪80年代为了方便交通,沿街的楼房才又再一次后退并且建成幸福大楼。至此火车站的站前广场变得更加开阔,但空间布局依然不明亮,几十年来一直保留着这种凑合、应付的设计。

火车站重建之后,对面区域绿树成荫,树冠相连形成巨大的"树屋顶",重新规划了空间格局。站前广场表层都覆盖了一种涂层,透气、透水性极大增强。

一块用混凝土浇筑而成的"地毯",上面有立体的不规则的圆形图案,与提契诺州的园林风格相互辉映。混凝土的表面由黑白两种颜色构成,如同片麻岩的表面结构。法国梧桐(悬铃木)是苏黎世的一种常见树木,在树干高5米的地方形成巨大的树冠。这两种不同的元素使得广场配置合理,并且整个造型简单明了。

霍斯特·博耐特与雕塑家协会共同设计了一个喷泉,喷泉由整块石板雕刻而成,主要采用的是从Cevion市运来的麻粒岩、花岗岩,设计参考了提契诺州的设计风格。一个由道格拉斯冷杉做成的直径8米的木环为乘客和居民提供了休憩之所。

景观设计中的瑞士印迹

UVEK Headquarters, Ittigen-Bern
伯尔尼伊蒂根 UVEK 总部

项目地点：伊蒂根市

景观工程师：Raderschall Landschaftsarchitekten AG, Meilen ZH

设计单位：GWJ Architekten AG, Bern

建筑工程师：Marchand und Partner AG, Bern

HLKKS 工程师：Enerconom AG, Bern

电气工程师：CSP Meyer AG, Bern

项目面积：约 13 800 平方米

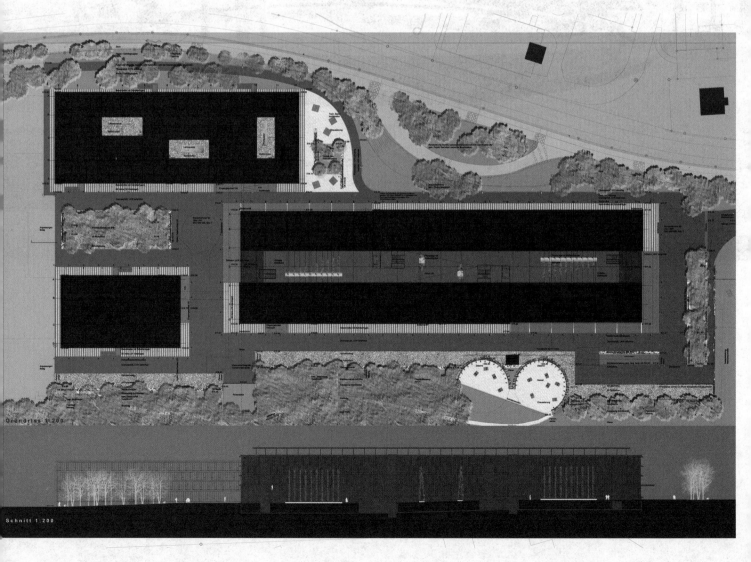

任务

在伊蒂根市磨坊大街2-6号原固瑞特-沃尔布拉股份公司的区域新建一个可以容纳约1100工位的行政中心。三座新建的大楼都是伊蒂根市交通部住宿区整体方案的一部分。议会曾于2002年批准了该民用项目的信贷。2004年春项目开始奠基，这是整个项目的第一步。第一期工程从2005年10月份开始到2006年3月份结束，包括交通部、民航局、能源部、公路部和空间发展部的建设。园林建筑的任务包括外部空间和内部庭院的设计。其中沃尔巴拉（Worbla）的建设是一个单独的项目，与上述工程同期施工。

外部空间

一些经典的方案使得外部空间具有清晰、结构分明和可持续发展的特点。大规模应用简单的建筑材料以与城市绝大部分的建筑形成统一协调的效果。

● 所有的沥青行车道边缘都没有磨边,而是铺设了粗糙且透水性强的涂层。

● 所有的停车带都用砾石铺置,旁边用钢制栏杆隔离。

● 建筑区域地面也用砾石铺设,有巨大缝隙的地方铺设混凝土板。

● 铺满碎石的格宾挡土墙是森林到建筑的过渡。

● 树林区种植不同的树种,对城市布局起到补充作用。

● 串联的照明装置和相应的灯光效果与刚性的设计效果相辉映。

内部庭院

三个内部的花园庭院用旧时代风格的植物景观进行装饰：原生态的天然材料，如石灰华和泥炭通过植物、水、空气和灯光的作用会发生一系列缓慢的变化。这种设计不是模仿传统的造园手法而是利用了栽培植物的一些自然现象。

三个内部庭院采用整体设计的方式，运用了相同的材料，这种材料和室外设计所用材料一致。碎石材料的每一个小立方体是由泥炭、黏土和层状凝灰岩板组成的。充足的水源和背阴创造了湿润的环境，从而为苔藓、草、蕨类和地衣等植物的生长提供了有利条件。泥土灰色的碎石上也因此布满了绿色的、新鲜的茸毛植被。气氛紧张的办公室就设置在园林区域的对面，能缓解紧张的办公氛围。

所使用的材料是运用硬水与空气的作用烧结而成的凝灰岩。这是一种多孔的石头，在烧制的时候也可以将树木或叶子包裹进去进行烧制。凝灰岩特别适合苔藓类植被在湿润环境中着床成长。黏土是曾经普遍使用的建筑材料，其特征是美丽的细腻颗粒形成不同的层次。但是黏土十分贫瘠，只适合少数植被生长。泥炭也是一种过去常用的建筑材料，在寒冷地区通常被当作燃料。在缺乏树木资源的冰岛，泥炭也被用作建筑材料。而在瑞士泥炭产生的沼泽是受到保护的。因此想在庭院里设计一个长满植被的泥炭沼泽在瑞士不是一件容易实现的事情。利用泥炭建立沼泽地的技术和知识，在中东地区很难找到。于是向冰岛国家博物馆寻求技术支持，他们熟悉如何在传统建筑中利用煤球建造沼泽地。其实最难的地方在于寻找原材料。最终，通过开垦的方法得到了泥炭。最后的施工是由马汀·劳赫完成的，他拥有丰富的建筑材料施工的经验。

长廊的混凝土地面上铺设了钢制的长管。顶层的天花板和钢制长管之间有开花类攀爬植物，气味芬芳。植物位于垂直方向，与内庭构成了三维的视觉效果。

City Center Leue, Männedorf
门内多夫城市中心

项目地点：门内多夫

设计单位：raderschallpartner AG Landschaftsarchitekten, Meilen

项目面积：4700平方米

社区教堂、市政大厅和商业区紧密相连,形成了城市自由空间区域,每个区域都有特定的功用和独特的氛围。人行专用区三面毗邻街道,与广场区融为一体。维斯中路延伸向西,并且建造了一个无台阶的步行区连接两个区域的道路。

在上述三座房子的中心区域是铺设好的地下车库,形成了心脏地带,而且很好地联系了这三个不同作用的区域。树木和水域区设计在老街道的北部,这里也是广场的起点。密集的带有伞状树冠的菩提树形成了巨大的树叶屋顶,广场上的一个巨大的天然石雕塑上面有水幕。市政大厅和社区教堂之间的楼梯通往地势较低的米特维斯广场。台阶高低有序,像梯田一般,为市民和游客提供了可以落脚歇息的地方。一棵盛开的樱花树正好为台阶区提供了树荫。

停车场的入口和大的分岔路口都分布在维斯中街上。一个宽的地槽和其他划定边缘的元素将楼房区和街道区分开。沿着老路和山路两侧种植着菩提树,这种树的身影遍布该地区的每个角落。

灯光

易乐广场夜晚通过吊灯实现照明,吊灯悬挂在绳子上,随风轻轻摆动,犹如浪涌一般。与此同时维斯广场和台阶区域也被柔和的灯光照亮,就像走进温馨的客厅一样。维斯中路和其他街道则通过传统的形似蜡烛的灯照亮。

标牌

为了方便区分不同的区域设计了专门的指示标牌。镀铬钢板上凸起的字母组成地名指示牌。通道和入口用不同高度的指示牌区分。

景观设计中的瑞士印迹

Klybeckquai, Basel
巴塞尔 Klybeck 码头

项目地点：巴塞尔

设计单位：Fontana Landschaftsarchitektur, Basel

瑞士莱茵口岸是瑞士唯一的国际港口，一直位于远离城市的地方。过渡期也需要一个过渡性质的解决方案：临时的海滨长廊早已对公众开放，但是复杂的港口物流要求岸上必须具备相应的配套设施。因此，协调利益和具体使用情况之间的关系成为该项设计的首要任务。与港口运行水岸之间的一块陆地上将1300米长的围栏最终削减到200米，这样岸边就空出90％左右的空间。

新的海滨长廊扮演着城市与工业区之间缓冲地的角色，并且符合港口的主题——存储、倒仓、堆积——这是港口的常用功能。岸上三条已经废弃的铁轨是设计中非常重要的一部分。它们被布以不同的模块元素。一条轨道被一种与水结合的材料覆盖，成为缓速形式的轴线。另外一条轨道以三种不同的元素交替布置，这三种元素可划分空间，并且具有从文化上区分空间的功能：圆盘形状的座椅与开放的轨道（覆盖自然植被或种植一年生植物）及工业金属集装箱是这三种主要元素。集装箱内的植物，以耐旱植物为首选，如杨树和桦树，桦树脚下一般生长着喷泉草，这是一种典型的园林植被，集装箱中的植被将港口区和外部空间有机联系起来。三种元素具有灵活性，可以根据不同的需求而随时进行改变。虽然可以随时改变，但美感却是永恒的，改变要时刻注重与美感相协调。

Voltamatte, Public Park, Basel
巴塞尔 Voltamatte 乐园

项目地点：巴塞尔

设计单位：Berchtold.Lenzin Landschaftsarchitekten, Liestal/ Zurich

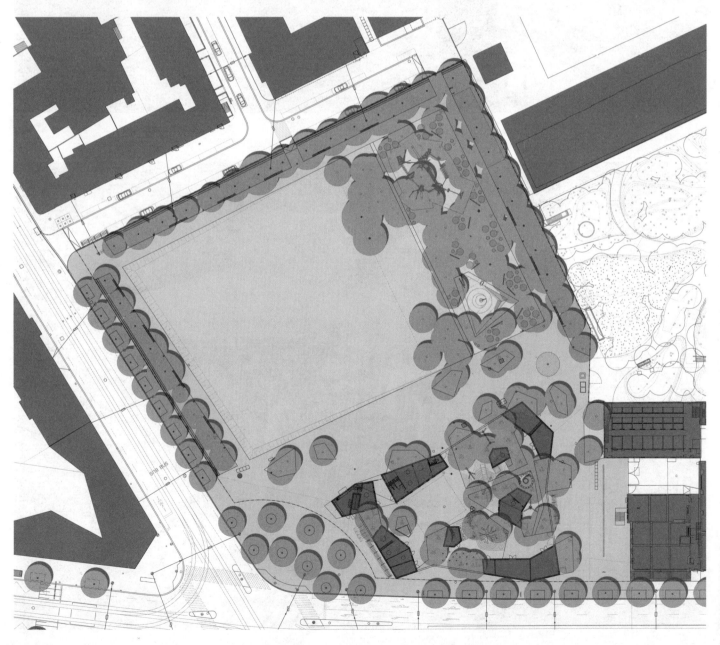

早期背景和政治进程

Voltamatte位于巴塞尔市的最西北端。在1862年的老地图上,Voltamatte就被标示成一个绿色空间,在圣约翰产业区的煤气厂和发电厂周围种植了绿树。1938年,一场激烈的公投运动后,创建了Voltamatte乐园,它是市区第一家乐园。目前公园的布局和设计可以追溯到当时流行的形式,1957年,Voltamatte乐园进行了扩建,修建了瑞士第一家冒险乐园。

从那以后,这座冒险乐园,就成了重要的露天活动场所,更为重要的是,它与这个有着众多移民人口的建筑密集区融为了一体。

地下交通网Nordtangente的建设,使得Voltamatte的部分地区可以安装公共设施,以及充当临时道路。地下交通网完工后的地面复原工程,被视为是改造整个建筑群,使之适应新环境,改变用途模式的一个机会。

在对面的Voltamatte俯瞰冒险乐园

设计的基础设施利于植物更强壮地生长

Voltamatte以开放的草地和橡树林为特色

在杂草丛生的区域,基础设施设置得很少,到目前为止只设置了一个篝火场地

柏油路将散布在碎石区域中的冒险亭台联系在一起

栽植的橡树周围,橡木干节段被水平放置可作为座椅,多年生草本植物在橡木林间构造出不同的功能区

位于橡木林中的儿童乐园内的北极熊雕塑格外引人注目

项目描述

对现存的公园首先进行了重新规划。现存的冒险乐园,由于通道受到限制,所以从东部移到了南部,到了一个更加开放的区域。沿着宁静的Voltastrasse,这个被围栏包围的冒险乐园,整个建筑群得到了明显的完善。这块区域在特定时间段只对14岁以下儿童开放,受到慈善协会员工的监督。各个现存的小型建筑物,作为整体的一部分,进行了重新改造。主建筑包括一个公共便利场所和一个Kindertankstelle(字面意思为:儿童加油站)。儿童加油站的服务群体是儿童以及陪伴的成人。除了出租游乐设施,运营商从公园部门将部分清洁职责转了过来。这确保了在夏季开放期间的有序管理。这个儿童加油站位于主建筑的西端,凭借处于公园和Voltaplatz中间的位置,以及各自的服务项目,其成为了一个交汇地。同时,标明了入口,使人回想起公园刚开始布局时就存在的柠檬树凉亭。

改造后在公园中央空出来一块地,设计成一个供人放松的绿荫空间。其内有一片小橡树林,随意栽种的植物具有多种功能,树干、滑梯、秋千和绳网组成小游乐场,另外还有由沙子和水构成的幼儿活动区。

Voltamatte是以草地得名(Matte就是草地的意思),因此,在中央地带,保留了一块面积以扩大草地。开阔的草地与周围斑驳的树木相互映衬,形成了一片趣味盎然的闲暇逗留区,供人们在此消磨时间。在冒险乐园新址与草地之间,有一块硬场地,安装有路灯,作为一块操场,可以用来举办多种活动。这个地方在公园中央,起着整合元素的功能。

一圈环形路环绕着草地,人们可以坐在路边的长凳上眺望草地、平地以及树林中的一切活动。可移动的公园设施允许游客按照自己的需求来进行布置。

随意设置的亭台楼阁聚集在一起,形成一个有保护的冒险村落,也是公园的南部边界

SWING Commercial Building, Walliseller
瓦利塞伦商业大楼

项目地点：瓦利塞伦
设计单位：KuhnLandschaftsarchitekten, Zurich
客户：Losinger Construction AG
建筑设计：B.E.R.G. Architekten / agps architecture

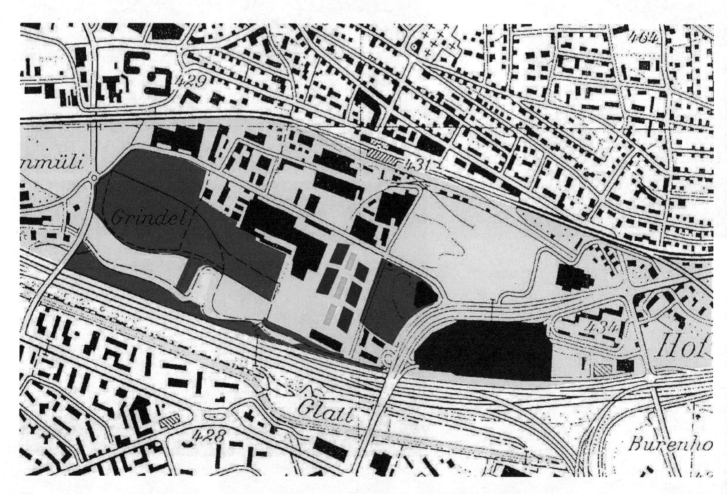

三个位于高层停车场的甲板花园顺势利用了山谷景观的美景。三个小小的、密布植物的世界隐居于大型建筑物为主的环境中。它们除了具有观赏作用之外，也具备实际功用：可作为进出的通道。

三个小花园都遵循相同的结构：地面层、灌木层、乔木层。不同的材料、不同的植被营造出不同的氛围。简单却集中，三个小花园在人造的周边环境中显得格外引人注目。

花园一

设计元素包括：随意布置的天然石地砖；景天（珊瑚红景天）、红豆杉（德国曼地亚红豆杉）、榛花（蜡瓣花属）、落羽杉（落羽杉属）、山樱桃（樱桃琼花）。

花园二

设计元素包括：河卵石，圆形混凝土板；蕨类植被（蹄盖蕨）、杜鹃（酒红杜鹃花）、白果（银杏）。

花园三

设计元素包括：混凝土板；羽毛草（极细针茅）、欧洲赤松（樟子松）、火枫（茶条槭）。

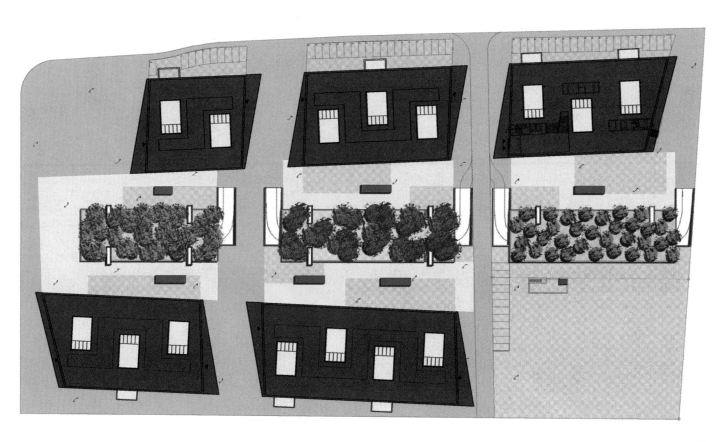

Cristal Park, Biel
比尔市克里斯特公园

项目地点：比尔市

设计单位：Klötzli Friedli Landschaftsarchitekten AG, Bern

工程商：Tschopp + Kohler Engineering

环境专家：Geotest, Zollikofen

摄影：Roger Grisiger photography

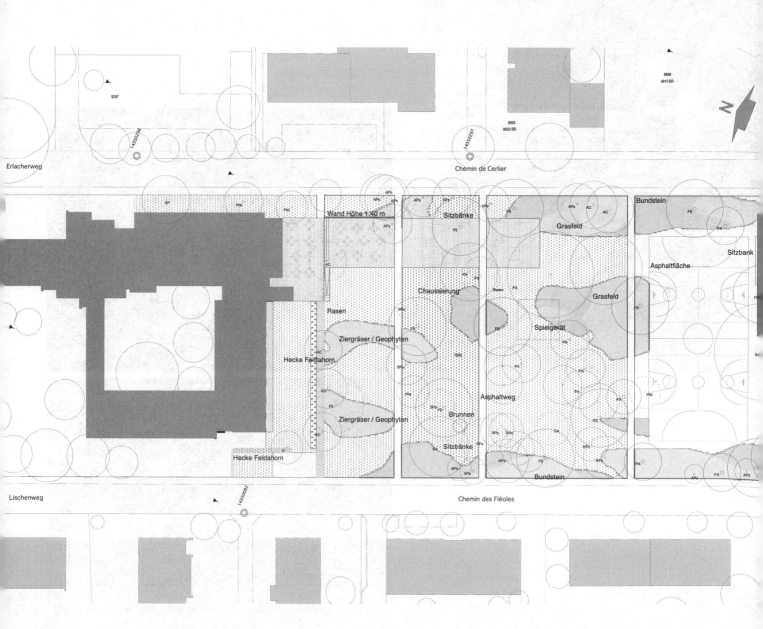

此前这块区域是一个垃圾处理场,因此人们认为不适合在上面兴修建筑物,而应在上面修建一座公园,于是,这片位于郊区的世外桃源才有了出现的可能。

因为这个原因,附近一块之前被指定为公园的区域被改造成了一个建筑区。新公园坐落在地区中心地带,不再是沿着街道的一个缓冲带。重新开发公园的主要目的是营造一片开阔地,满足各行各业的当地居民的各种需求。

2006年,比尔市为此专门举办了一届景观设计大赛。

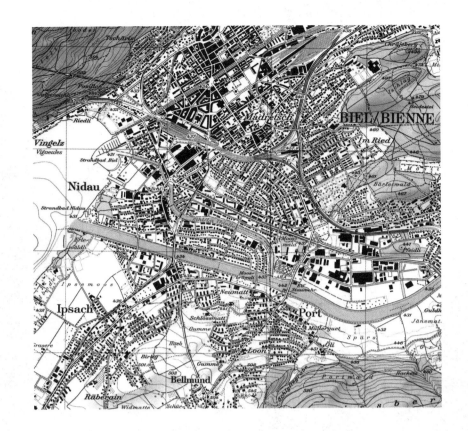

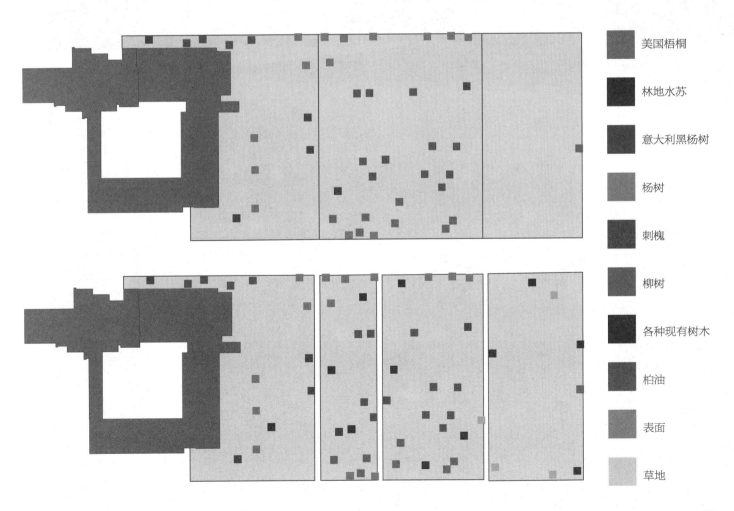

美国梧桐

林地水苏

意大利黑杨树

杨树

刺槐

柳树

各种现有树木

柏油

表面

草地

现有结构

原来的开阔地有三个区域：老年公寓前的一片开阔草地，中心地带的一片树木密集区，末端的一片沙砾区。这个结构被保留了下来，只做了适当调整，补充了一块沙砾地。铺了柏油路，将各个部分相隔与连通。

树概念

很大程度上保留了原有的树木，被集合到新的概念中。

这种植物结构主导着新的布局原则。在公园中心，银色的落叶树搭配了经典的黑色叶子的公园树（山毛榉，Swat Magret）。很显然，在"世俗化"的先锋树种与"高贵"的公园树种之间，在亮与暗之间，存在着互动。

为了使人回想起树间不规则生长的小草，沿着边界，以及在公园内部地带栽种了新的地杨梅，在春天还会种上白色的郁金香，进一步为公园带来生气。

改造前

改造后

公园

这是一个不显山露水却又充满意义的方案：

原有的区域和树木被保留了下来，只在需要的地方做了完善。因为这个原因，从一开始，公园就被当地居民所接受，并且有着多重用途。原来的步行小道铺上了柏油，除了人可以行走外，轮椅和婴儿车也可以在上面推行。

一块水泥地面的区域和一块操场形成了新的玩耍区。周围的长凳可供人们休息放松，不同年龄的人们可在此交流沟通，公园附近就是老年公寓。

THE SWISS TOUCH IN LANDSCAPE ARCHITECTURE
景观设计中的瑞士印迹

Enlargement to Accomodate Muslim Graves, Rosenberg Cemetery, Winterthur
温特图尔罗森博格公墓穆斯林墓区扩建

项目地点： 温特图尔市

设计单位： Berchtold.Lenzin Landschaftsarchitekten, Liestal/ Zurich

委托方： 温特图尔市政园林部门

项目面积： 3700 平方米

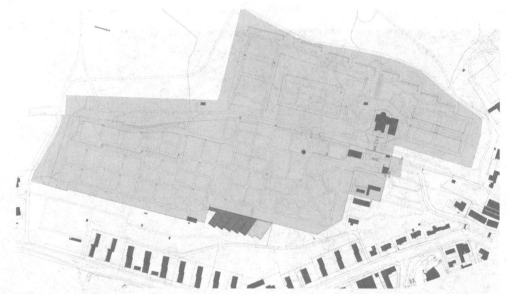

Rosenberg公墓及穆斯林墓区周界

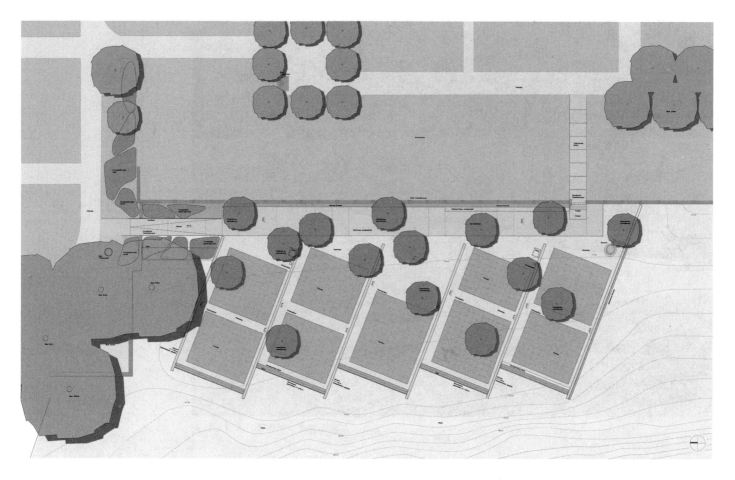

背景

温特图尔市位于苏黎世西北部,是瑞士第六大城市,人口超过10万。最近几十年来,温特图尔市由工业城市逐步转变为服务、教育和文化中心,市内有众多博物馆及其他休闲设施。自2006年至2010年,议会制定了使移民人口更好地融入温特图尔市这一目标。其中的一项举措就是使不同宗教团体的已故成员可以安葬在温特图尔市。穆斯林团体占该市总人口还不足12%。

2006年8月,温特图尔伊斯兰青年协会向苏黎世伊斯兰组织协会(VIOZ)递交了一项超过300人签字的申请,要求在温特图尔设立伊斯兰墓地。最重要的一项要求是按照伊斯兰独特的丧葬形式举行葬礼以及墓地须朝向麦加圣地。当时,已经从瑞士其他大城市吸取了有关穆斯林墓地建设的经验。

原有公墓穆斯林墓区扩建项目入口处概览

抛光石灰石混凝土喷泉

原有公墓混凝土墙与扩建项目之间的多年生草本植物

参考摩尔人建筑模型设计的喷泉，溢出型盆槽以及敞开式导流槽

在温特图尔墓地建设可行性研究中，证实了可以通过扩建罗森博格公墓把穆斯林墓区纳入其中。在遗产保护方面，为了找到最佳解决途径，2008年举行了一场竞赛，邀请参与方递交项目方案。由市政府人员、瑞士少数民族协会会长、苏黎世伊斯兰组织协会副主席、温特图尔市土耳其协会委员会的一名成员以及温特图尔市伊斯兰教－阿尔巴尼亚协会教长组成项目小组，负责相关准备和评判工作。为了协助陪审团做决定，还邀请了来自景观设计和建筑设计领域的专家担任评审官。陪审团最终推荐了最能反映当事人诉求的作品作为获奖作品。

残疾人友好型坡道入口,植有修剪灌木

由于受全球金融危机的影响,进一步的规划被无限期推迟。但是,在2010年底,温特图尔市议会获得了信贷支持计划。该项目获得了来自各方的支持,使它得以顺利开展。

项目描述

由于罗森博格公墓是市重点保护遗址,扩建项目要与当前项目实现无缝对接,以便使该处历史遗迹原貌保持不变。扩建场地紧邻当前墓地,确保了可以通过其他地块和一条小路使其与相邻的日常生活地区分开。扩建项目与相邻的建筑区交织在一起,成了综合景观的一部分。由于仅仅采取了很少的景观措施,它可以与斜坡地形融为一体。墓区设有残疾人友好型通道,可以通过两个路口进入宽阔的水泥路,并间隔栽种小树。水泥路连通各墓地。1.2米高的粗面混凝土石灰石隔离出各个单独墓地。

石灰墙全部朝向麦加圣地，确保了墓地的方向也朝向麦加圣地。墓区的标志性特征为开阔的拉马克唐棣树林。这种植物有多重特点：风景如画、暗色树皮、白色花朵以及在秋天迷人的颜色。石灰墙末端和直角处种植的低矮树篱把墓区与邻近田园区分开。墓地被砾石路面环绕并种植有各种不同颜色的、类似地毯的百里香。灵活调配的椅子可供人们使用。

鉴于现有公墓的典型特征，在水泥路的南部新建了一个抛光石灰石混凝土喷泉。涓涓的水声、溢出型设计的盆槽，以及敞开式导流槽参考了摩尔人建筑模型。作为伊斯兰葬礼文化的重要元素，在中心位置设置了供送葬者在遗体下葬之前吊念死者的设施。在单独墓地方面，五个墓区具有高度的适应性。整个布局设有218个成人墓地、111个儿童墓地和73个私人墓地，再次扩充后，可以再容纳160个简单墓地。按照法律规定，墓地必须在至少二十年内保持原状，在这个条件下，每个墓地可以使用三次。整个扩建项目的特征为间隔良好的拉马克唐棣和粗面石灰石混凝土墙。

墓地周围各种颜色、类似地毯的百里香

从外面看位于周围草地之上的独立石灰岩混凝土墙

THE SWISS TOUCH IN LANDSCAPE ARCHITECTURE 景观设计中的瑞士印迹

Re-naturalisation of the River Aire, Geneva
日内瓦亚耳河的重新规划

项目地点：日内瓦

设计单位：Atelier Descombes Rampini sa, Geneva

项目面积：600 000平方米

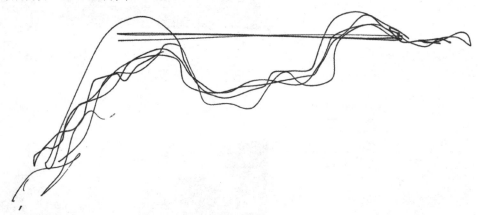

沿着亚耳河的地面修整工程从19世纪末就开始了，包括意义重大的农业灌溉工程和河床挖深工程，1945年工程才告一段落。亚耳河复兴工程自从2001年以来一直是一项日内瓦水道重新规划的工程，代表了亚耳河两岸地貌和景观的重构。从一开始，就寻求在农业生产、城市发展和旅游胜地的必要性与适当扩展以及连贯的自然空间重构之间找到新的平衡。此项工程恢复了几乎已完全被遗忘的景观特征，比如排水沟、树篱和沼泽。人行道的平整，以及运河的重建和河床的疏通，保证了水的合理流动和动植物群的连续性，地貌和居民的安全因此得到了维护，也考虑到了人们进行水上活动的可能性，从而能够与多样化的自然区域和谐相处。

| THE SWISS TOUCH IN LANDSCAPE ARCHITECTURE | 景观设计中的瑞士印迹

Cassarate River Delta, Lugano
卢加诺 Cassarate 三角洲治理方案

项目地点：卢加诺

设计师：Sophie Agata Ambroise Officina del Paesaggio

顾问：土木工程/Passera 和伙伴工程公司、生物学家/Luca Paltrinieri、地质学家/Urs Luechinger

项目成本：593 万瑞士法郎

图片所有：archivio Officina del Paesaggio

项目面积：10 000 平方米

图片所有：manuelabieri.ch

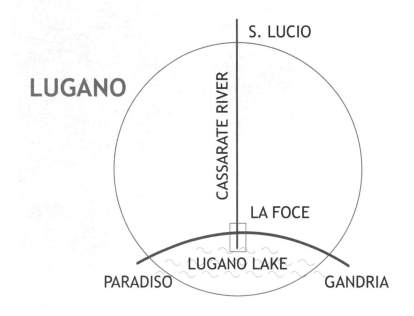
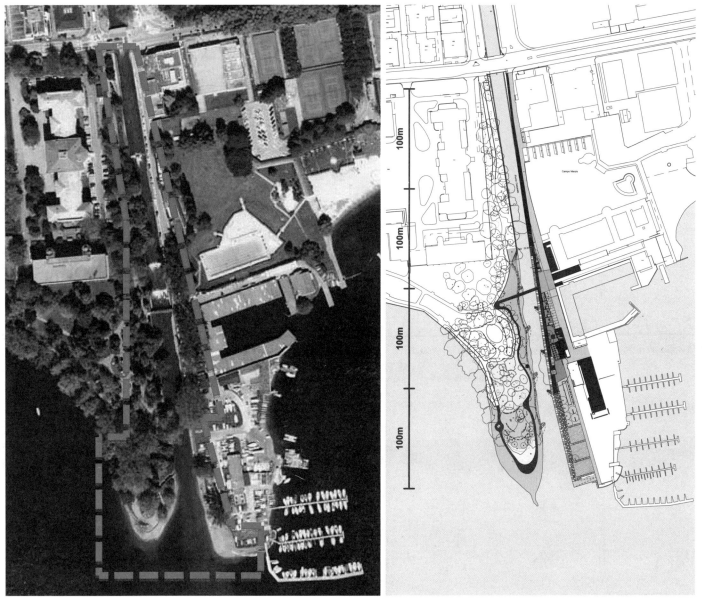

before the project（图片所有：archivio officina del paesaggio）

200

卢加诺位于瑞士南部的Ceresio湖上，Cassarate河被认为是城市的边缘。

为了繁荣河三角洲地区，卢加诺市政府2004年在提契诺洲和瑞士联邦的资助下举行了景观建筑设计竞赛。

该项目设计公司的设计理念强调河三角道是新卢加诺城市的中心，考虑将其作为城市设施的一部分。

未来卢加诺市会繁荣发展，这种发展将与城市的区域和积水盆地相结合，其将成为发展的主要载体。

这样一个新景观和自然的轴心，将推动许多极具娱乐性和开放性活动的开展。

项目最初的想法不是为了避免洪水的危险而关闭河道，而是为了创造两个新的河堤，人们可以在这两个新河堤上行走。为了市民能够再次靠近河道，在河道的右岸设置了一个流线形木质走道；河道的左岸，被再次设计为人造河堤，有娱乐设施和大石阶，是一个城市化的河岸。

2014年6月该项目落成，其非常有吸引力并受到市民的极大欢迎，卢加诺的市民可以在这个特色鲜明、河湖美景环绕的地方尽情享受。

after the project（图片所有：archigraphie.ch）

Museum Park, Kalkriese
卡尔克里泽考古公园

项目地点：卡尔克里泽

景观设计：Studio Vulkan Landschaftsarchitektur, Zurich

项目面积：200 000 平方米

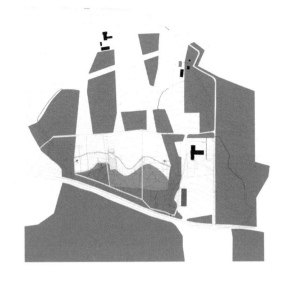

几个世纪以来,人们一直在寻找Teutonic森林,根据罗马作家Tacitus的作品,这片森林的所在地就是公元9世纪日耳曼人打败Publius Quintus Varus的罗马军队的地方。

这场大败使得罗马在莱茵河东岸的扩张停了下来。经过大范围的挖掘,现在证明,这个地方就是古战场的历史遗址。博物馆就坐落在27千米范围内的古战场的中心,占地200 000平方米。collin of Kalkriese驻扎在南边,有日耳曼人修建的堡垒确保安全,北部的沼泽地形成了一条自然的狭长地带,在几千米长的战线上,罗马军队有被奇袭的危险。

公园设计的关注点是:历史景观是博物馆的真实展示。在南部植林,砍掉遗址开阔地带原先不存在的森林,公园又回到了战争时的原始模样。

罗马军团的行进路线由铁板标示了出来,同时用铁柱来标示由日耳曼人修建的高墙的原有高度。不同的时间层次包括了过去和现在,意味着今天的农业用地及道路,都是公园的一部分。参考今天历史的教学法,考古公园不仅是历史物品的简单重建,也是对历史事件和历史环境的抽象描述。

West Axis, Zurich
苏黎世西部项目

项目地点：苏黎世西部

设计单位：asp Landschaftsarchitekten AG, Zurich

　　　　　Stücheli Architekten, Zurich

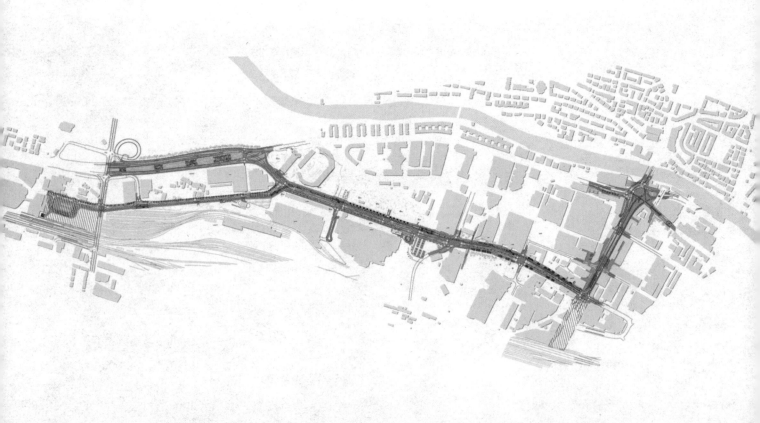

苏黎世西部曾经是一片工业区，是该市重要的发展区之一。西阿斯特可以作为这个区域的交通中轴线。这里有一条可以通往城市主要道路的高速公路，新建的体育场和苏黎世艺术大学催生了新的人行道和相关设施的建立，住宅和商业建筑需要在大区域内实现小的功能分区。在苏黎世西区铺设新的电车线路要求道路空间也需做出相应调整。该设计意图实现高品质功能分区的功用，因而进行了有针对性的设计，使得每个区域都有不同的作用。运用特殊的配色方案，使得电车的基础设施设备像具有地区特色的家具一样。在绿化方面，根据不同的街道布局选择不同种类和不同生长形态的树木。这两个设计使得西阿斯特地区结构分区合理，并且让开车不再枯燥："绿色的长城"在建筑物前随着季节的变化而变化，春天是细腻柔和的绿色，而秋天是显眼的红色，色彩的变化让城区充满活力。夜晚的灯光设计生动活泼。在高速公路并入伯尔纳大街的地方是一块特别显眼的绿色区域，这是城镇的入口：白杨和松树形成一个巨大的交通环岛。环岛内部苔藓、地衣、草丛和花卉随季节的变化而变化，形成五颜六色的"地毯"。但是这个6000平方米大的"公园"是不允许进入的。这是对城市密度的讽刺———一片就在眼前的绿洲，却只为沥青"沙漠"中生存的植物和动物而设置。

THE SWISS TOUCH IN LANDSCAPE ARCHITECTURE　景观设计中的瑞士印迹

East Village, Stratford (former Athletes Village 2012), London
伦敦东村

项目地点：伦敦

设计公司：VOGT Landschaftsarchitekten, Zurich

客户：Lend Lease

设计团队：Fletcher Priest Architects, Arup, Biodiversity by Design, Speirs and Major Associates, David Bonnett Associates, Tim O'Hare Associates, Waterwise Solutions, BMT Fluid Mechanics, Gardiner & Theobald, RPS Planning / Quod Planning

项目面积：开放空间 150 000 平方米

作为在前斯特拉特福德铁路区的一个开发和城市改造工程，于2002年初步构思形成了斯特拉特福德市开发方案，这片铁路区位于东伦敦斯特拉特福德市中心的北部。

伦敦取得2012年奥运会和残奥会的举办权后，东村并入斯特拉特福德市主规划之中。由于要举办奥运会和残奥会，使得开发时间从原来的15～20年，不得不缩短，以便能够为奥运会和残奥会的17 000名运动员提供住宿。

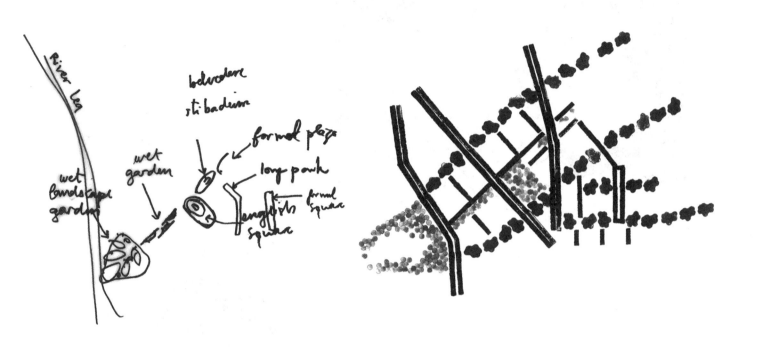

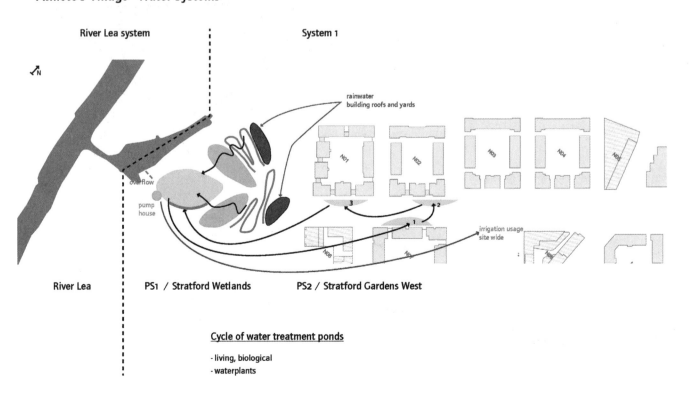

东村2012年的按时交付有着不小压力，因此设计团队的首要关注点是考虑2012年奥运会后社区的开发问题。2012年后社区包括居住区、零售区、办公楼和教育设施。

东村位于西部利河与东部现存雷敦居民区之间。其自然特点和周围的城市环境一直作为包括街景在内的开放空间设计的景观参考。

周围地形、原生植物和水道都为公共空间、广场和街景环境的特征提供了素材，营造出一处具有鲜明特点的地方。除了加强与周围环境的联系外，设计的主要原则是参考英国景观花园的传统风格来开发景观，并开发成一个水可持续利用的工程化景观。开发这样一个景观，不仅要营造一种地域感，并且要实现最小化灌溉，重复利用排水，创造地形多样化以及结构和栖息地多样化，在整个计划和整个湿地环境中，广泛运用原生和本地植物，这一切都需要景观建筑商、生态学家和工程人员之间的紧密合作。团队曾经的技术挑战成了这片公共领域设计的灵感来源，为东村一系列的设计战略提供了资料。

这个方案的目标在于营造一个诗意化的基础设施景观。

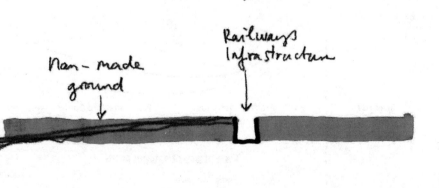

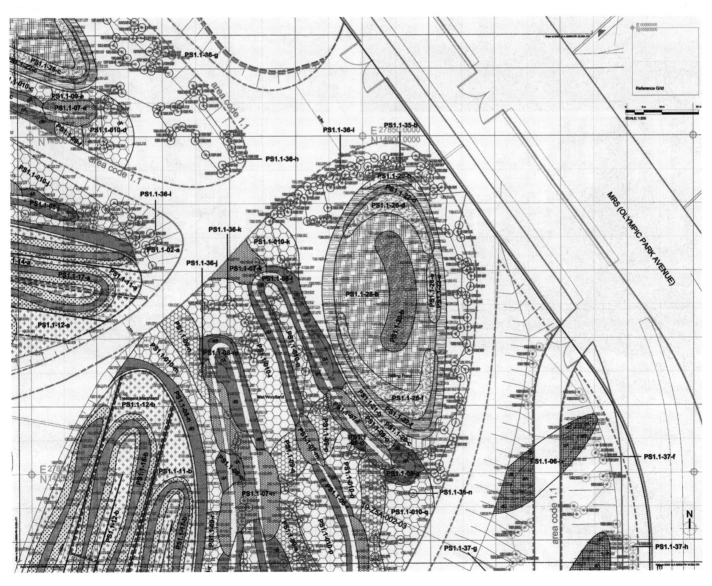

Route de Meyrin, CERN, Geneva
日内瓦 CERN 梅林大道

项目地点：日内瓦
设计单位：Studio Bürgi, Camorino
项目面积：60 000 平方米

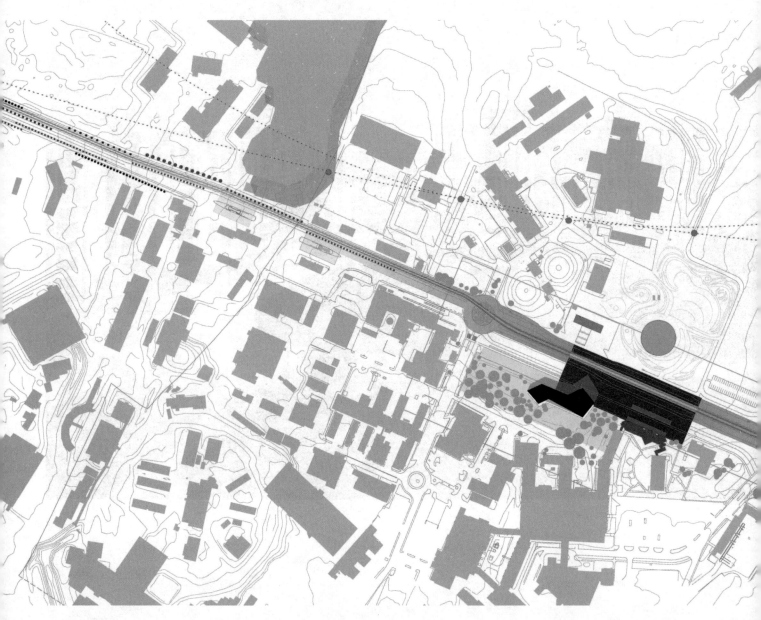

随着梅林大道这条CERN（欧洲核子研究组织）的A出口和B出口之间的道路建筑设计竞赛获胜者的宣布，CERN与世界的联通迈出了重要一步。获胜的项目冠名为"Metaphoros"，由Paolo Bürgi of Ticino工作室选送。

项目正式开工还需一些时间，但是想要在2014年看到通往CERN大门的样子，可以参观在GLOBE开幕式上举办的展览，展会持续到1月28日。展览聚焦在获胜作品上，但是也为第二名的作品提供了一席之地，即CERN天宇环，一个环GLOBE的建筑及景观建议，这个作品将在接下来的一个阶段即CERN公共空间的重新开发过程中，与Metaphoros无缝对接，而且这项工程需要外部投资。从长远来看，前景规划是将A和B出口之间的整个区域开发成一个充满生气的CERN和社会的接口，将一座新的主建筑与一个容纳1000人的大礼堂结合起来，这个大礼堂将和邻区共用。

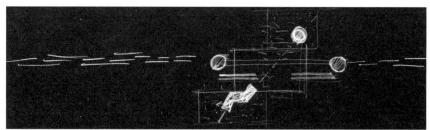

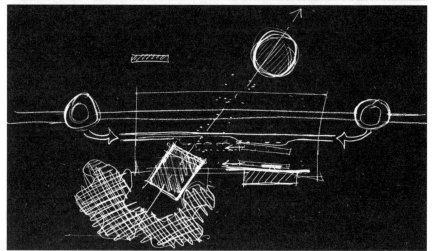

与社会的衔接是CERN所有事务中最重要的一个部分，主区为CERN提供了一个特别重要的地方。梅林大道的设计大赛就出于这样一种与社会衔接的理念而组织的，作为CERN、Republic和Canton of Geneva之间的一个合作，合作的这三方将为项目提供资金。大赛也得到了Ville de Meyrin和瑞士联邦政府的支持。起初就宣布了这是包括日内瓦、沃州和邻邦法国的整个地区的更加广泛的协调开发。

那么，从局部来说，Metaphoros就是CERN向社会开放，与更广泛的公众分享新思维的标志。在更广泛的程度上，也完美地隐喻了CERN是社会不可分割的一部分，以及其拥有的对科学的全球化视角。

CERN 标志性的接待区，是该地区的有力象征，增强了CERN的国际地位。

获胜作品通过4个关键概念重新设计了CERN的入口区：

● 一个宽敞的步行广场，将GLOBE与未来的主建筑连通；

● 沿着梅林大道，会员国的旗帜迎风飘扬；

● 一块巨大屏幕，实时显示CERN的科学发展成果；

● 一个更加便捷的轨道交通服务。

通过这种方式，这项工程体现了CERN国际化、科学化和现代化的精神。

LA PLACE DES PARTICULES

ARRIVE AU CERN APRES AVOIR PARCOURU LA RUE AVEC DES RANGEES D'AR
PLACE OUVERTE, UNE SURFACE DELIMITEE ET CLAIREMENT RECONNAISSAB
LIEU PRIVILEGIE, TRAVERSE PAR LES PIETONS ET AVEC UN TRAFIC RALENTI.
LE NOUVEAU BATIMENT POURRA DEVENIR UN ELEMENT DE CONNEXION ET D

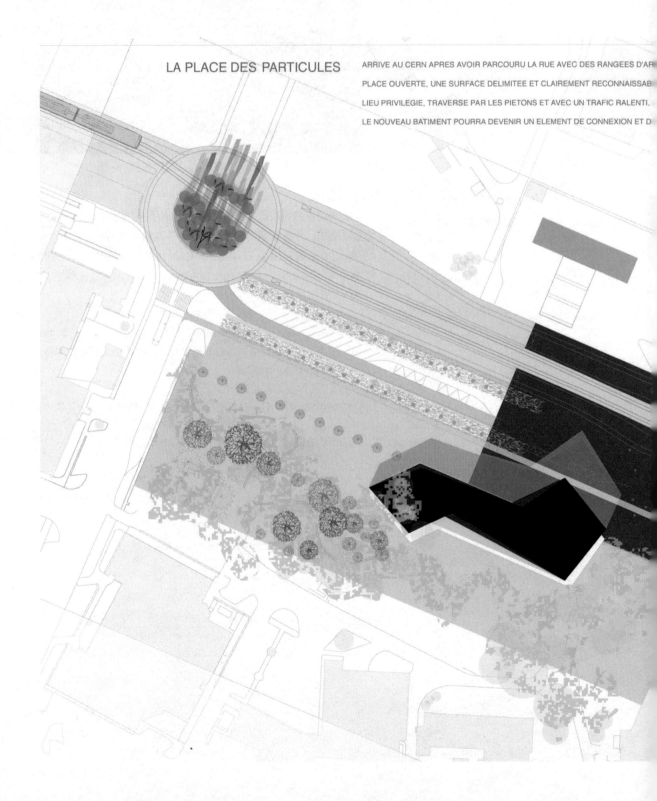

SANCE PYRAMIDALE, VOUS RENCONTREZ UN PREMIER POINT SIGNIFICATIF: LA PORTE D'ACCES, UN IMPORTANT ESPACE VERT A TRAVERSER. VOUS ATTEIGNEZ L'ESPACE CENTRAL, UNE
T MET EN RELATION ENTRE EUX LES PRINCIPAUX BATIMENTS PUBLICS REPRESENTATIFS ET D'ACCUEIL AU CERN. LA GRANDE PLACE, REHAUSEE A UN NIVEAU UNIQUE, DEVIENT AINSI UN
ENDRA UNE PLATEFORME OUVERTE AUX RENCONTRES, AUX RELATIONS SOCIALES, A TOUTES SORTES D'EVENEMENTS D'EVOCATION OU DE CELEBRATION.
TION ENTRE LE GLOBE, LA PLACE, E LE "PARC DU MUSEE", SITUE A L'ARRIERE, QUI INVITE LES VISITEURS ET LES COLLABORATEURS DU CERN A UN MOMENT DE REPOS A L'OMBRE.

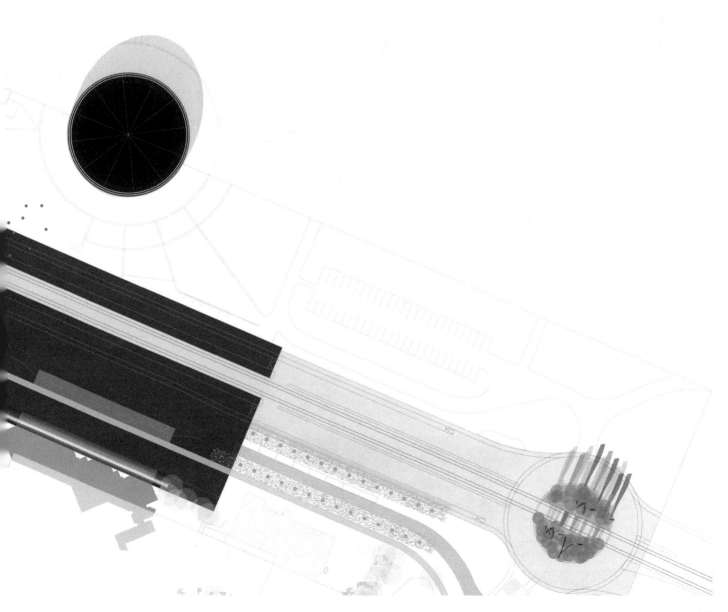

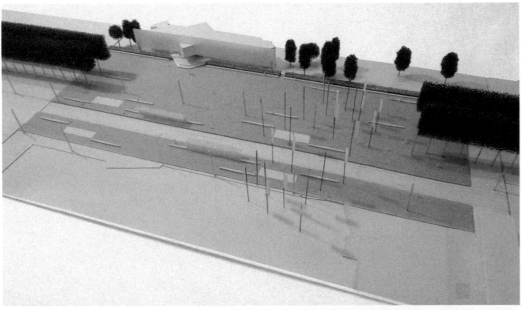
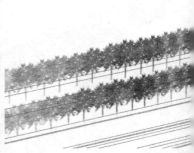

Fensterfabrik G. Baumgartner AG, Hagendorn-Cham
鲍姆加特股份公司窗户工厂的厂房扩建项目

设计地点：卢采恩

设计单位：koepflipartner landschaftsarchitekten, Lucerne;
　　　　　Graber and Steiger, Lucerne (architects)

完成时间：2014年

规划面积：16 500平方米

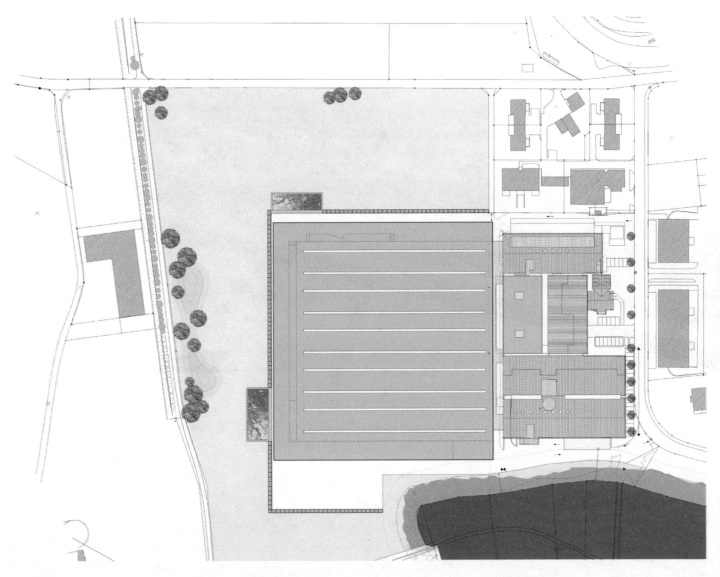

起始状况

为扩建窗户工厂，2002年就开始竞赛，以获得一个能综合考虑现有景观特点的解决方案。景观通过农业处理、地块的划分、灌木和森林的利用痕迹，给人留下深刻的印象。而从远处看，各个部分以及蜿蜒穿过平面的Lorze河的河道清晰可见。这是一种因人为干预而产生的景观。据考证，该区域有人居住的历史已有2000多年，窗户工厂区域发掘出的一个罗马人使用过的磨坊就是最好的证明。同时这也是一种始终处于变化之中的景观。历史证明其就是窗户工厂周围二战期间建造的、用于代替湿草甸而获得农业生产面积的排水沟。这种湿草甸过去对农民来说是一种劣势条件，但现在其对稀有植物和动物来说是撤退区域，如在楚格州其他的地方一样，在Hagendorn很少能见到这样的场景了。也就是说，Lorze河在楚格湖Alpenblick区域或在Reusspitz汇入Reuss河。

理念

我们对厂房进行了扩建，屋顶的打造作为回归Hagendorn景观的唯一机会。其目标是在屋顶建设一个紫色沼泽草小型草场。紫色沼泽草小型草场是一种湿草甸类型，常见于阿尔卑斯山山前地带，厂房扩建区域、排水沟施工前或许占主导地位的是小型草场类。一个屋顶项目的意义在于，这种紫色沼泽草小型草场所展现的是一种人工的小型草场类。如果不定期修剪和维护，将成为灌木，进而演化成林地。为此作为滞留盆的屋顶得以形成。

项目

鉴于条件及知识所限，目前尚无法实现在屋顶建设一个真正的小型草场，尤其无法在16 500平方米的面积上来完成。在规划中我们没有标准或经验可循。为估算涉及的土壤混合物、种子等的风险，我们一共设置了8个采用不同基质和种子构成的实验场。2004年设置，随后两年进行了评估。一种由47.5%当地产生的挖掘材料、47.5%可循环的屋顶废砖和5%沸石组成的基质混合物，或者说一种土壤改良基质，经两年的观察证明其效果是最好的。屋顶的密封这样形成，即可持久堵住5厘米的水，最多能堵20厘米的水。更多的水则通过屋顶的溢水管流出。

发展

播种后,在屋顶上面将不采取浇水、除草等园艺辅助措施。其目的是让植物群自然发展,成为稳定的植物群落。7年后一个湿润的小型草场就形成了。在这个湿润的小型草场中,干燥阶段也能如有水阶段一样,交容易地克服困难。而且紫色沼泽草,就是小型草场的指示植物,能缓慢地生长起来。但如先前一样,单独的新种可由境外迁入或重新消失,并如一个天然小型草场一样,苔藓层能缓冲水分平衡。

小型草场的维护费用在屋顶的建设中占有一定的比例,但与天然草场不相上下。不需要的桦木或牧场树芽必须去掉。

前景和可能性

屋顶面积的扩大不仅为植物群也为动物群提供了新的可能性。不仅是植物，动物们也需要更大的、相关联的小型草场以便生存。在规划屋顶时曾经有人表达了这样的想法：或许凤头麦鸡也应该出现在屋顶上。凤头麦鸡是Riedwiesen的一种典型的"居民"，直到1980年其还能在Hagendorn周围的景观中被经常见到。但现在却很少能见到，几乎已消失。具体原因有很多，但适宜的群落生境的破坏、农业生产中土壤经机械处理后所存在的危险等都是其中的原因。屋顶的扩大将大大的吸引凤头麦鸡它的到来。这里为幼鸟的培育提供保护的条件。但吸引凤头麦鸡只是扩大屋顶生物种群的一个例子。事实上，甲壳虫、蜘蛛、蝴蝶等都能在这里找到群落生境。显然，屋顶参与了周围景观的生物节律。

Novartis Campus Park, Basel
巴塞尔诺华公司园区

项目地点：巴塞尔

设计公司：VOGT Landschaftsarchitekten, Zurich

客户：诺华制药公司（巴塞尔）

总体规划：Vittorio Magnago Lampugnani Studio di Architettura, Milano

项目面积：63 000 平方米

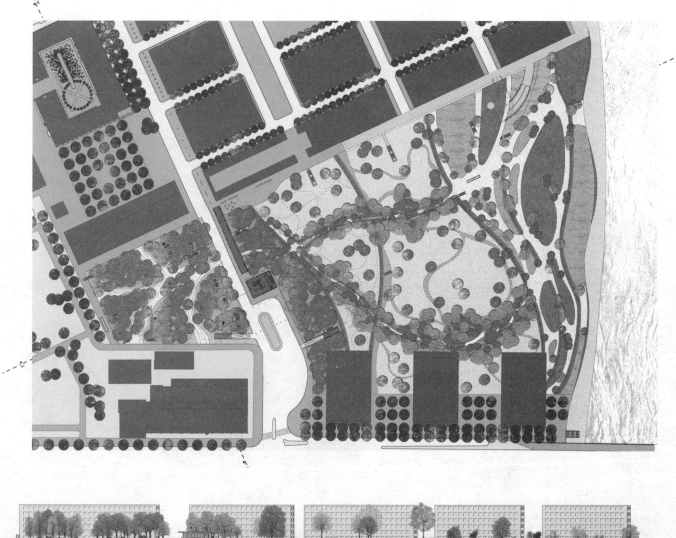

诺华园区位于巴塞尔的边境三角地带，它是诺华·圣·约翰工业园区的一部分。

它之前是一个生产制造园区，如今被改造成一个研究和行政管理中心。这一园区不仅是诺华公司的总部所在地，它也应当成为一个吸引员工和具有各种功能的园区，让员工在这里会集和工作。众多的办公大楼，由不同的建筑师设计，如今它们有的已经处于施工阶段，有的还处于设计阶段。

该园区三面都有新建筑，而另一面则面向莱茵河，且有悬臂式的台阶。在园区有多层的地下车库。除了莱茵河畔，没有其他的文化或自然类型来帮助构建该项目的设计理念。就这一地点而言，莱茵河是一个重要的元素，它具有强有力的影响力，因此莱茵河是设计理念的重要来源。

但是，巴塞尔的地图上显示出一个鲜为人知的地貌。在城市结构的下方，我们可以看到莱茵河谷上游的地势，它由冰川和沉积岩所构成（是史前时期的见证）。在这里，我们看到一个具有独特特征的地貌——由冰川和水侵蚀所形成：冰川造就了平原和盆地地貌；融水构成了冰川盆地和峡谷；水平的融水沉积，创造了冰川地貌和冰川堆石，以及冰川漂砾（由于冰川运动而留下来的巨大岩石）。这一进程已经上演了成千上万年，如今依然尚未停息，不断地塑造和重塑莱茵河中上游河谷和其地貌。这一进程，可以顺着河流而上，追溯到它的源头——阿尔卑斯山。

从巴塞尔附近河谷洼地的截面图可以看出，有一系列清晰的地下岩层，最古老的岩层，距离河床最远。根据这一地质结构的模拟，可以发现自然植被也遵循着一个相应的时间系统。这一冰川谷地的形成，对冲积土地上的植被而言，是最理想的生长之地，因为它们喜欢潮湿环境。由于免受水流和洪灾的侵袭，这些植被在这里扎根并慢慢生长，最后在谷地边缘构成林地。它们依然是这一地貌冰川历史的见证。在巴塞尔附近的森林中，在橡树和山毛榉中间，有很多巨大的岩石。

诺华园区的设计，主要取源于这一"隐形的地貌"，它是莱茵河上游谷地地形和植被现象的组成部分。周边的自然地貌在小规模上得以重塑，并通过设计最终形成一个具有特定氛围的公园地貌。这一园区可以理解为莱茵河上游谷地的一个地貌的横切片。它有三个部分组成。

园区的区域，位于河岸距离最远的地方，被设计为一个森林，里边有当地的木本植物，而且还有巨大的冰川漂砾，它们依然是冰川活动最古老的见证。其中的道路，犹如雪崩时的样子，其直接通向莱茵河，并且两旁有巨石。

园区的中间部分，由大面积的草坪所构成，其中种植了各种各样且具有异域特色的木本植物。在这里，主干道分叉为两个小道，通向两个不同的方向，它们穿越不同的地势以及茂密的植被。

在第三区域，毗邻莱茵河的低地势向外扩展开来，有广阔的水域和植被地带。植被地带（从冲击区域到卵石海岸）主要由高的沼泽植物所组成，几乎没有任何树木。

THE SWISS TOUCH IN LANDSCAPE ARCHITECTURE　景观设计中的瑞士印迹

Alp Transit-Depot, Sigirino
Alp Transit（穿越阿尔卑斯）工程：Sigirino 火车站

项目地点：卢加诺

设计公司：Atelier Girot GmbH, with ITC & IFEC Eng.

客户：Alp Transit San Gottardo

预算：2000 万欧元

地面面积：19 公顷

体积：370 万立方米

该项目是修建高速隧道，把意大利和北部欧洲连接起来。在其中一个隧道工程中，将会从Sotto Ceneri山脉挖掘出370万立方米的石头材料，并运输和堆砌在Sigirino火车站。这座火车站，位于卢加诺北边5千米的位置，对面就是Sigirino小村庄（紧邻古老的St. Gotthard公路）。这一地方毗邻一个重要的自然保护区，从公路主干道和连接瑞士与意大利的火车线路上，就可以看到该景色。由于其地理位置十分特殊，大量的石头材料将会堆砌在此，以及工程致力于保护旅游业和环境质量的承诺，客户认为：从一开始，就需要一个景观建筑师参与到设计团队之中。火车站分台阶的多面地形结构，提供了收集水资源的系统，并为游客提供了参观小路。鉴于该工程规模浩大、过程复杂，3D的地形形态学上的分析以及可视化技术在整个过程正得到了广泛的应用。基于政府规划，这一地点的火车站，将会成为重要的交通枢纽。依据3D模式，制作动画片，以便帮助理解原地的体积规模。项目将于2020年竣工。

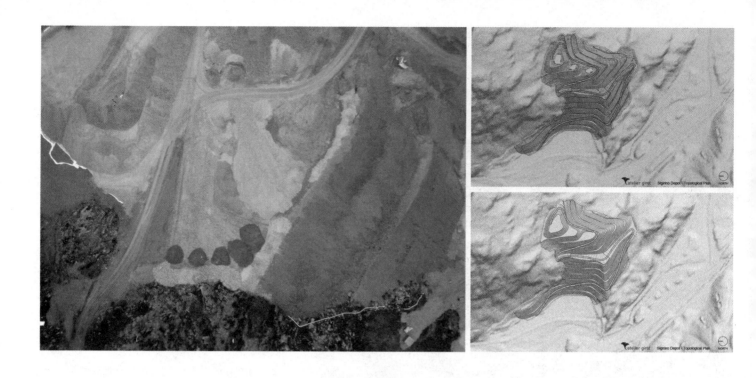
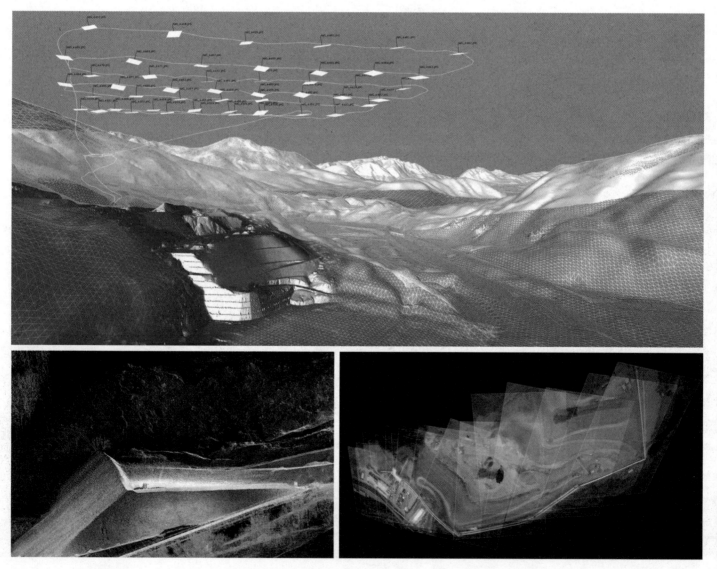

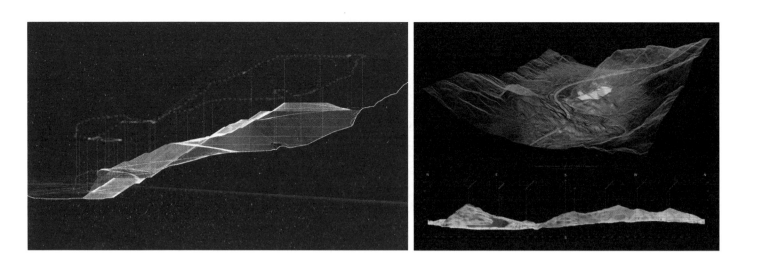

Reconsidering a Mountain, Orselina-Cardada
奥尔塞利纳—卡尔达达缆车改造

项目地点：洛迦诺市

设计单位：Studio Paolo Bürgi

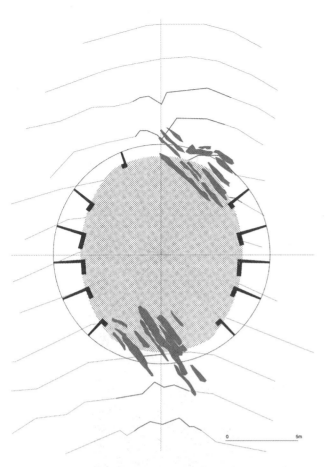
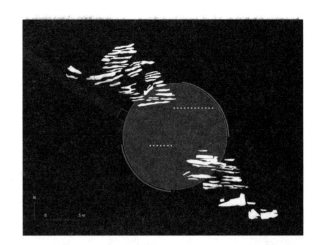
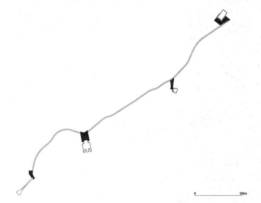

奥尔塞利纳 – 卡尔达达缆车的改造项目，使人们重新发现一片广阔且风景优美的山区。由于现代人们的生活日趋分散化，所以能够找到一处令人震撼和陶醉的风景，这样的机会越来越少。从这一角度而言，卡尔达达的风景，主要在于它最本质的文化方面，即重新阐释这一地区的历史底蕴。这是一次重新评估，源于对当代影响的认识。本项目通过一些规模较小但具有重大意义的人工干预措施而构成，它能够为观赏者提供新的有利位置，让他们重新发现这一地区独特的自然本质。站在一个由钢和钛筑成的巨大网状建筑上，观赏者可以远眺马焦雷湖；由混凝土结构筑成的平台（契梅塔地质观测台）讲述着当地的地质历史；它们与周围的环境完美地融为一体。

图书在版编目（CIP）数据

景观设计中的瑞士印迹 /（瑞士）雅各布主编；翟俊译 . -- 南京：江苏凤凰科学技术出版社，2015.3
ISBN 978-7-5537-3829-1

Ⅰ. ①景　Ⅱ. ①雅　②翟　Ⅲ. ①景观设计－作品集－瑞士－现代　Ⅳ. ① TU986.2

中国版本图书馆 CIP 数据核字 (2014) 第 212799 号

景观设计中的瑞士印迹

主　　　编	[瑞士] 迈克尔·雅各布（Michael Jakob）
译　　　者	翟　俊
项 目 策 划	凤凰空间/高雅婷　李　妍
责 任 编 辑	刘屹立
特 约 编 辑	陈丽新
出 版 发 行	凤凰出版传媒股份有限公司
	江苏凤凰科学技术出版社
出版社地址	南京市湖南路1号A楼，邮编：210009
出版社网址	http://www.pspress.cn
总　经　销	天津凤凰空间文化传媒有限公司
总经销网址	http://www.ifengspace.cn
经　　　销	全国新华书店
印　　　刷	利丰雅高印刷（深圳）有限公司
开　　　本	889 mm×1 194 mm　1/16
印　　　张	17
字　　　数	228 000
版　　　次	2015年3月第1版
印　　　次	2024年10月第2次印刷
标 准 书 号	ISBN 978-7-5537-3829-1
定　　　价	278.00元

图书如有印装质量问题，可随时向销售部调换（电话：022-87893668）。